24.00

Astronomy and Astrophysics Series Volume 12

Radiation Transfer and Stellar Atmospheres

Pachart

Astronomy and Astrophysics Series
General Editor: A.G. Pacholczyk

Volume 1: T.L. Swihart
Basic Physics of Stellar Atmospheres

Volume 2: T.L. Swihart
Physics of Stellar Interiors

Volume 3: R.J. Weymann, T.L. Swihart, R.E. Williams, W.J. Cocke, A.G. Pacholczyk, J.E. Felten
Lecture Notes on Introductory Theoretical Astrophysics

Volume 4: E.R. Craine
A Handbook of Quaisistellar and BL Lacertae Objects
(Reference Works in Astronomy)

Volume 5: A.G. Pacholczyk
A Handbook of Radio Sources. Part I, Strong Extragalactic Sources, 0-11 Hours. (Reference Works in Astronomy)

Volume 6: V.N. Zharkov, V.P. Trubitsyn
Edited by W.B. Hubbard
Physics of Planetary Interiors

Volume 7: G.W. Collins, II
The Virial Theorem in Stellar Astrophysics

Volume 8: G. Rossano and E.R. Craine
Near Infrared Photographic Sky Survey:
Field Index
(Reference Works in Astronomy)

Volume 9: M. Heller and D.J. Raine
The Relativity of Space-Time

Volume 10: G.R. Gisler and E.D. Friel
Index of Galaxy Spectra
(Reference Works in Astronomy)

Volume 11: Z. Alksne and J. Ikaunieks
Edited by J.H. Baumert
Carbon Stars

Volume 12: T. L. Swihart
Radiation Transfer and Stellar Atmospheres

Radiation Transfer

and Stellar Atmospheres

Thomas L. Swihart
University of Arizona

Pachart Publishing House
Tucson

Copyright © 1981 by the Pachart Corporation

No part of this book may be reproduced by any mechanical, photographic, or electronic process, or in the form of a phonographic recording, nor may it be stored in a retrieval system, transmitted, or otherwise copied for public or private use without written permission from the publisher.

Library of Congress Catalog Card Number: 80-83367
International Standard Book Number: 0-912918-18-7

Pachart Publishing House
Box 35549
Tucson, Arizona   85740

Table of Contents

General References   xi

Chapter 1. The Equation of Transfer   1

    1. Intensity and Derived Quantities   1
    2. The Absorption Coefficient   6
    3. The Emission Coefficient   8
    4. The Equation of Transfer   9
    5. The Source Function   11
    6. Special Integrals for Plane Media   19

Chapter 2. The Gray Atmosphere   23

    7. Introduction   23
    8. The Eddington Approximation   26
    9. The Method of Discrete Ordinates   29
   10. Isotropic Scattering   36

Chapter 3. The Non-Gray Atmosphere   39

   11. The Model Atmosphere   39
   12. Reduction to the Gray Solution   45
   13. Corrections to the Temperature Distribution   49
   14. Convection   52
   15. Semi-Empirical Models   60
   16. Continuous Absorption and Blanketing   62

Chapter 4. Line Formation   65

   17. Line Absorption and Emission   65
   18. The Two-Level Atom   68
   19. Line Broadening   74
   20. Profiles, Equivalent Widths, Curves of Growth   85

Chapter 5. Polarization 97

    21. Pure Polarized Radiation 97
    22. General Polarized Radiation 101
    23. Transfer in a Magnetic Plasma 107
    24. Rayleigh Scattering and the Sunlit Sky 112

Appendixes

    1. Physical and Astronomical Constants 121
    2. Problems 123

Index 129

List of Figures and Tables

Figure 1. Defining intensity, 2
Figure 2. Geometry of the measurement of stellar flux, 5
Figure 3. Geometry in a plane atmosphere, 19
Figure 4. Observed and gray model solar fluxes, 37
Figure 5. Effects of convection on the temperature distribution, 59
Figure 6. The polarization ellipse, 98
Figure 7. Intensity and degree of polarization of the daylight sky, 116
Figure 8. Scattering geometry, 118

Table 1. Temperatures in exact gray and the Eddington approximation atmospheres, 29
Table 2. Gaussian weights and points, 31
Table 3. Integration constants in gray problem, 35
Table 4. Exact values of $q(\tau)$, 35
Table 5. Model solar atmosphere, 51
Table 6. Types of pressure broadening, 79

Preface

This is a revised and expanded version of the author's Basic Physics of Stellar Atmospheres, published in 1971. The change in title reflects the fact that stellar atmospheres are no longer the exclusive subject.

The main addition is a new chapter on the transfer of the polarization properties of radiation. Polarization is an increasingly important part of modern astrophysics, and a brief introduction to it in the context of radiation transfer and stellar atmospheres does not seem inappropriate.

The entire text has been rewritten. Chapters 2 and 3 contain more numerical examples. New problems have been added in Appendix 2, and some answers and hints for finding solutions have been included. It should also be noted that the symbol F now stands for the full flux. I was never able to see much justification for using πF for the flux, but old traditions die hard.

The purpose of the book has not changed. It is intended for those who prefer a sort of extended summary rather than the great detail that any subject has to offer if the student really wants to find it. Because of this the book has been deliberately kept short, and the attempt was constantly made to keep mathematical accuracy from obscuring physical concepts.

Some would prefer to see more prominence given to the most recent research problems and the latest numerical data. Before these can be appreciated, however, the student must have a grasp of the basics. I do not intend a book that is long and complete enough to do an adequate job on both. Of course, the student who wishes to carry out research in this field must go well beyond the material presented here.

I appreciate the comments and suggestions made by Prof. A. G. Pacholczyk; and the help of Mrs. Maxine Howlett made this work much easier.

                                        Thomas L. Swihart

Steward Observatory
University of Arizona
Tucson, Arizona
May 15, 1980

General References

Aller, L. H. The Atmospheres of the Sun and Stars, 2nd ed., Ronald, 1963.
Carbon, D. F. Model Atmospheres for Intermediate- and Late-Type stars, Ann. Rev. Astron. Astroph. 17, 513, 1979.
Chandrasekhar, S. Radiative Transfer, Oxford, 1950. Reprinted by Dover, 1960.
Gibson, E. G. The Quiet Sun, NASA SP-303, 1973.
Gray, D. F. The Observation and Analysis of Stellar Photospheres, Wiley, 1976.
Greenstein, J. L., ed. Stellar Atmospheres, Chicago, 1960.
Jefferies, J. T. Spectral Line Formation, Blaisdell, 1968.
Mihalas, D. Stellar Atmospheres, 2nd ed., Freeman, 1978.
Pecker, J.-C. Model Atmospheres, Ann. Rev. Astron. Astroph. 3, 135, 1965.
Unsöld, A. Physik der Sternatmosphären, 2nd ed., Springer, 1955.

Chapter 1: The Equation of Transfer

1. Intensity and Derived Quantities

The basic unit of radiative transport is the intensity. The intensity of a radiation field can be defined in the following manner. Consider the two elements of area $d\sigma$ and $d\sigma'$, as illustrated in Figure 1. These areas have normals along the unit vectors $\hat{n}$ and $\hat{n}'$, respectively. We wish to determine the radiation energy which crosses $d\sigma$ within the time interval dt and which later crosses $d\sigma'$. We assume for the present that there is no matter present to absorb or emit radiation, so we are concerned only with the geometry of the situation. The energy in question is apparently proportional to the time interval dt as well as the projected area ($d\sigma \cos \theta$), where θ is the angle between the direction of propagation and the unit normal $\hat{n}$. We also expect the energy transferred to be proportional to $d\omega$, the solid angle of $d\sigma'$ as seen from $d\sigma$. The factor of proportionality in this is the intensity I:

$$dE = I \cos \theta \, d\sigma d\omega dt \qquad (1.1)$$

Intensity is energy per area per solid angle per time, and it is a function of position, direction, and time. It is also a function of the frequency or wavelength of the radiation. One can define a monochromatic intensity per frequency interval I_ν such that $I_\nu d\nu$ is the intensity due to all waves with frequencies between ν and $\nu+d\nu$. There is a similar definition of I_λ, the intensity per wavelength interval. The total or integrated intensity is then given by

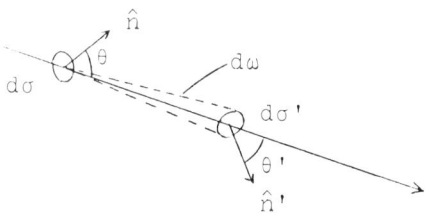

Figure 1. Defining Intensity

$$I = \int_0^\infty I_\nu d\nu = \int_0^\infty I_\lambda d\lambda \tag{1.2}$$

Frequency dependence is not important in the present section, and it will not be explicitly indicated. The relations considered here are valid for both monochromatic and integrated quantities.

An important quantity derived from the intensity is the mean intensity J. This is the intensity averaged over all directions. Since the total solid angle around a point is 4π steradians, the mean intensity is given by

$$J = \frac{1}{4\pi} \int I \, d\omega = \frac{1}{4\pi} \int_0^{2\pi} \int_0^\pi I(\theta,\phi) \sin\theta \, d\theta d\phi \tag{1.3}$$

In this expression θ and ϕ are the usual spherical angles.

Another quantity of importance which is derived from the intensity is the flux $\vec{F}$. The arrow indicates that flux is a vector quantity, although we will usually be concerned with only the magnitude of the flux F. Let $\hat{s}(\theta,\phi)$ be a vector of unit length in the direction specified by the spherical angles θ and ϕ. Then the flux at any point is given by the integral

$$\vec{F} = \int I(\theta,\phi) \, \hat{s}(\theta,\phi) \, d\omega \tag{1.4}$$

If $\hat{n}$ is a unit vector in a fixed direction, the component of the flux along $\hat{n}$ is found by taking the scalar product:

$$F(\hat{n}) = \vec{F} \cdot \hat{n} = \int I \, \hat{n} \cdot \hat{s} \, d\omega \tag{1.5}$$

By choosing the direction of $\hat{n}$ to define $\theta = 0$, one has $\hat{n} \cdot \hat{s} = \cos\theta$. Equation (1.5) then takes the familiar form

$$F(\hat{n}) = \int I \cos\theta \, d\omega \qquad (1.6)$$

A comparison of equations (1.1) and (1.6) indicates that the component of flux in a given direction is the net energy per unit area and time crossing a surface normal to that direction. Unless otherwise stated, the term flux will hereafter refer to the magnitude of the vector flux. The vector flux is the Poynting vector $(c/4\pi)\ \vec{E} \times \vec{H}$ of electromagnetic theory.

By making the substitution $\mu = \cos\theta$, we can write the flux as follows:

$$\begin{aligned} F &= \int_0^{2\pi}\int_0^{\pi} I(\theta,\phi) \cos\theta \sin\theta \, d\theta d\phi \\ &= \int_0^{2\pi}\int_{-1}^{+1} I(\mu,\phi)\mu \, d\mu d\phi \\ &= \int_0^{2\pi} d\phi \left[\int_0^{+1} I(\mu,\phi)\mu \, d\mu - \int_0^{+1} I(-\mu,\phi)\mu \, d\mu \right] \qquad (1.7) \end{aligned}$$

The first term in equation (1.7) is due to radiation being propagated in the outward or positive hemisphere ($\mu > 0$), while the second term represents radiation traveling in the negative direction ($\mu < 0$). If these partial fluxes into the respective hemispheres are denoted by F^+ and F^-, then

$$F = F^+ - F^- \qquad (1.8)$$

The total or net flux is the excess of that in the positive direction over that in the negative direction. In an isotropic radiation field, the partial fluxes in all directions are equal and the net flux is zero: there is no net transfer of energy in any direction.

Equation (1.1) gives the energy in a given pencil of radiation, i.e., the energy crossing a given surface element in time interval dt and confined to a given elemental solid angle. This energy is contained in the volume ($d\sigma \cos\theta \, cdt$), where c is the speed of light; therefore, the radiation in that particular solid angle contributes an amount ($I \, d\omega/c$) to the energy per unit volume at the point considered. The total energy density is then given by

$$u = \frac{1}{c} \int I \, d\omega = \frac{4\pi}{c} J \qquad (1.9)$$

The energy density is proportional to the mean intensity J.

Radiation of energy dE carries momentum dE/c in the direction of propagation. According to equation (1.1), the given pencil of radiation contributes the amount (I $\cos^2\theta$ × $d\sigma d\omega dt/c$) to the normal component of momentum carried across the surface $d\sigma$ in time dt. This quantity, taken per unit area and time, is the pencil's contribution to the scalar pressure at the given point. Thus the radiation pressure is

$$P_r = \frac{1}{c} \int I \cos^2\theta \, d\omega \qquad (1.10)$$

It should be emphasized that the radiation quantities considered here are defined for any point, not for an extended surface or volume. The flux is the ratio of energy to area, taken in the limit as the area goes to zero. The other quantities are defined in an analogous way.

The intensity of radiation is not affected by the geometry of the region of interest, but only by the physical addition or subtraction of energy through emission and absorption. To see this, consider again Figure 1. The energy dE as given by equation (1.1) is that which first passes through $d\sigma$, then later passes through $d\sigma'$. We could also consider the energy dE' which is passing through $d\sigma'$ and which had previously passed through $d\sigma$. The latter quantity is given by

$$dE' = I'\cos\theta' \, d\sigma' d\omega' dt \qquad (1.11)$$

The primed quantities are measured at a point on $d\sigma'$, and $d\omega'$ is the solid angle of $d\sigma$ as seen from $d\sigma'$. Since solid angle is projected area divided by the square of the distance, we have

$$d\omega = \frac{d\sigma' \cos\theta'}{s^2} \qquad d\omega' = \frac{d\sigma \cos\theta}{s^2} \qquad (1.12)$$

Here s is the distance between the two surfaces. It is apparent that dE = dE', and it follows that I = I'. Thus the intensity of a beam of radiation is not changed as it propagates from one point to another, except for sources and sinks along the way.

The intensity of starlight as seen from the Earth is the same as seen at the surface of the star, except for the amounts blocked off by the material between us and the star. The difference is that near the star it subtends a much larger solid angle than as seen from far away; flux and mean intensity, which are both integrals of intensity over solid angle, do fall off with distance from the source. Physical effects do not depend directly upon intensity. Excitation of atoms, for example, depends on the mean intensity, while producing an image on a photographic plate depends on the flux.

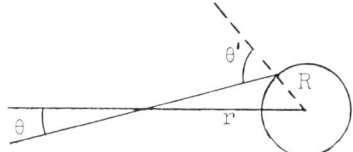

Figure 2. Geometry of the measurement of stellar flux.

Consider a point at distance r from a spherical star of radius R, illustrated in Figure 2. The flux from the star is

$$F = \int_0^{2\pi} \int_0^{\theta_0} I(\theta,\phi) \cos\theta \sin\theta \, d\theta d\phi \qquad (1.13)$$

where $\theta_0 = \sin^{-1}(R/r)$ is the angular radius of the star. Let θ' be the angle at the star between the normal to the surface and the direction to the point of observation. Then for neighboring points on the stellar surface,

$$r^2 \cos\theta \sin\theta \, d\theta = R^2 \cos\theta' \sin\theta' d\theta' \qquad (1.14)$$

As θ varies from zero to θ_0, θ' goes from zero to $\pi/2$. Then if the intensity can be expressed as a unique function of θ', the flux in equation (1.13) becomes

$$F = \frac{2\pi R^2}{r^2} \int_0^{\frac{1}{2}\pi} I(\theta') \cos\theta' \sin\theta' d\theta' \qquad (1.15)$$

Since the above integral is independent of r, the flux is seen to vary as r^{-2}. In the special case for which I is independent of θ', the flux reduces to $(\pi I R^2/r^2)$.

Note that it is not necessarily true that the intensity is a unique function of θ'. For a star that is spotted or nonspherical, for example, it is not true, and the flux close to the star does not fall off exactly as the square of the distance.

2. The Absorption Coefficient

This section is concerned with a description of how matter interacts with radiation to subtract energy from the radiation stream. Our interest is phenomenological; we are not concerned with the quantum mechanics of how to calculate accurate cross sections. Since frequency dependence is quite important here, the frequency subscript will be explicitly indicated for quantities whose dimensions include per unit of frequency interval. It will not be included on quantities such as the absorption coefficient which simply have a dependence on frequency.

The rate at which matter causes energy to be lost from a beam of radiation is described in terms of an absorption coefficient. A pencil of radiation of intensity I_ν will lose by absorption the intensity $dI_\nu(abs)$ on traveling the distance ds. The absorption coefficient is defined by the relations

$$dI_\nu(abs) = I_\nu k ds = I_\nu \kappa \rho ds \qquad (2.1)$$

where k is the volume absorption coefficient of dimensions (l^{-1}), κ is the mass absorption coefficient of dimensions (l^2/m), and ρ is the mass density of the material. k and κ are alternate quantities to describe the same thing. In practice one uses the form of the absorption coefficient that is most convenient, and this may be either the mass or the volume coefficient. Both will be used in the present work.

Absorption as described here includes any process in which a photon is removed from a pencil of radiation. If a photon is deflected into a new direction without being otherwise changed, the process is called scattering. Non-scattering refers to processes in which the photon is changed into other forms of energy, perhaps to reappear later as another photon. Absorption includes both scattering and non-scattering.

Let $W(s)$ be the probability that a photon will travel the distance s without being absorbed. Then $W(s_1+s_2) = W(s_1) \times W(s_2)$. With $s_1 = s$ and $s_2 = ds$, one finds, using a first order Taylor expansion,

$$W(s+ds) = W(s) W(ds) = W(s) + \frac{dW}{ds} ds \qquad (2.2)$$

According to equation (2.1), (kds) is the probability that

absorption will take place within ds, so (1 - kds) is the probability that it will not, which is W(ds). If this result is put into equation (2.2) and the integration carried out with k assumed constant, we find

$$W(s) = e^{-ks} \qquad (2.3)$$

Now let P(s)ds be the probability that absorption will take place between s and (s+ds). Then

$$P(s)ds = W(s)\{1 - W(ds)\}$$
$$= e^{-ks} kds \qquad (2.4)$$

The mean free path L is the average distance a photon travels before being absorbed. From the above we find

$$L = \int_0^\infty sP(s)ds = \frac{1}{k} \qquad (2.5)$$

The volume absorption coefficient k is the reciprocal of the photon mean free path. If k does vary with position, the integrals leading to equations (2.3) and (2.4) cannot be carried out, but equation (2.5) still defines the local value of the mean free path.

k and κ are macroscopic coefficients in that they describe the properties of matter in bulk. Absorption can also be described on the microscopic scale. Suppose that each absorbing particle has an effective cross section area of α as seen by photons of a given frequency. If there are N of these absorbers per unit volume, then a column of area A and length ds will contain (NAds) of the particles. A photon traveling along this column will see these (NAds) particles presenting a total area of (NAαds), which is the fraction (Nαds) of the area of the column. In other words, (Nαds) is the probability of absorption on traveling the distance ds. Comparing this with equation (2.1), we see that

$$k = N\alpha \qquad (2.6)$$

Thus k is the total cross section of the absorbing particles contained in unit volume, and hence the name volume absorption coefficient. Likewise, κ is the total cross section of unit mass of the absorbers.

It is apparent that the volume absorption coefficients

are additive, at least as long as the particles are far enough apart that they can act independently. If there are different kinds of particles in a mixture, having coefficients k_1, k_2, etc., then the total coefficient is $k = k_1 + k_2 + \ldots$. The same is not true of the mass absorption coefficients, unless the individual coefficients are given per unit mass of the mixture.

3. The Emission Coefficient

The radiation energy emitted by the volume element dV into the solid angle $d\omega$ in time interval dt and within the frequency range $d\nu$ is given in terms of a volume emission coefficient j_ν:

$$dE_\nu = j_\nu \, dV d\omega d\nu dt \tag{3.1}$$

The frequency subscript indicates that both j_ν and E_ν are measured per unit frequency interval. As with the absorption coefficient, the emission coefficient can be expressed per unit volume or per unit mass. If ε_ν is the mass emission coefficient, then $j_\nu dV = \varepsilon_\nu dm$, or $j_\nu = \rho \varepsilon_\nu$.

Just as the absorption coefficient includes all effects which subtract energy from a given pencil of radiation, the emission coefficient includes all effects which add to the energy. This includes radiation which is simply scattered from other directions into the direction of interest. This scattered energy must be added to whatever other forms of energy are converted to radiation and emitted by the volume element.

Consider the volume element dV as being the cylinder of cross section $d\sigma$ and of length dx. According to equation (3.1), the contribution of dV to the energy in the pencil of radiation is $(j_\nu \, d\sigma dx d\omega \nu dt)$. But $dx = ds \cos \theta$, where θ is the angle between the axis of the cylinder and the direction of propagation, and ds is the propagation distance within the cylinder. If we compare this with equation (1.1), we see that the intensity is increased by the amount

$$dI_\nu(em) = j_\nu \, ds \tag{3.2}$$

The contribution of any small region to the intensity is the volume emission coefficient times the size of the region in the direction of propagation.

4. The Equation of Transfer

When radiation propagates a distance ds, the intensity will be increased by the emissions and decreased by the absorptions along the path. Combining equations (2.1) and (3.2), we find

$$\frac{dI_\nu}{ds} = j_\nu - kI_\nu \tag{4.1}$$

It is understood that ds is measured along the path of propagation. Although intensity is a function of position, time, and direction, the direction is held fixed as one follows the radiation. Thus the dI_ν and the ds in equation (4.1) are related through $dI_\nu = (\partial I_\nu/\partial s)ds + (\partial I_\nu/\partial t)dt$. But ds is the distance the radiation travels in time dt, i.e., dt=ds/c. Equation (4.1) can then be written

$$\frac{1}{c}\frac{\partial I_\nu}{\partial t} + \frac{\partial I_\nu}{\partial s} = j_\nu - kI_\nu \tag{4.2}$$

It is only rarely that the explicit time dependence of the intensity need be taken into account. This does not mean that time plays no role, but it does mean that the physical quantities generally do not vary appreciably in the time it takes the radiation to cross the system of interest. When this is the case, we can set $\partial I_\nu/\partial t = 0$, and any time dependence is brought in through the variations in the absorption and emission coefficients. The derivative with respect to position then becomes total. Hereafter this condition will be assumed.

The distance element ds can be related to any desired coordinate system. If q_i are the coordinates of interest, then

$$\frac{d}{ds} = \sum_i \frac{\partial q_i}{\partial s}\frac{\partial}{\partial q_i} \tag{4.3}$$

In a cartesian system, dx = lds, dy = mds, dz = nds, where (l,m,n) are the direction cosines of the propagation direction. Then

$$\frac{dI_\nu}{ds} = \left(l\frac{\partial}{\partial x} + m\frac{\partial}{\partial y} + n\frac{\partial}{\partial z}\right)I_\nu \tag{4.4}$$

In a system with spherical symmetry, let r be the distance from the center and θ the angle measured from the radial direction. Then dr = cos θ ds, and rdθ = -sin θ ds, so

$$\frac{dI_\nu}{ds} = \left(\cos\theta\frac{\partial}{\partial r} - \frac{\sin\theta}{r}\frac{\partial}{\partial \theta}\right)I_\nu = \left(\mu\frac{\partial}{\partial r} + \frac{1-\mu^2}{r}\frac{\partial}{\partial \mu}\right)I_\nu \qquad (4.5)$$

On the right side of equation (4.5) the substitution μ = cos θ has been made. In a similar way one can write the equation of transfer in any coordinate system. Of course one will choose that system which takes greatest advantage of whatever symmetry the problem of interest possesses.

Multiply equation (4.1) by the following factor:

$$\exp\{\int_0^s k(s')ds'\}$$

The two terms involving the intensity then combine to make a perfect derivative, and the result is easily integrated:

$$I_\nu(s_2) = I_\nu(s_1)\exp-(\int_{s_1}^{s_2} kds') + \int_{s_1}^{s_2} j_\nu \exp-(\int_s^{s_2} kds')\,ds \qquad (4.6)$$

This equation gives the intensity at a given point in terms of properties of the same pencil of radiation at previous points along its path of propagation. The intensity at point s_2 is equal to that at any previous point s_1, attenuated by the exponential factor which represents the absorption between s_1 and s_2; to this is added all of the emission between the same points, again attenuated by the absorption between the integration point and s_2. If there is no absorption, the intensity increase is simply the sum of the emissions over the path, as equation (3.2) indicates. If there is no emission, the intensity is cut down by the exponential absorption consistent with equation (2.1).

If the explicit time dependence is kept, as in equation (4.2), then the solution is still of the same form as equation (4.6); the only difference is that each quantity in the expression is to be evaluated at its retarded time argument, i.e., at the light travel time interval in the past.

A simple and instructive case is the solution for a homogeneous medium in which the absorption and emission coefficients do not depend upon position. Then equation (4.6) is seen to reduce to

$$I_\nu(s_2) = I_\nu(s_1) e^{-k(s_2-s_1)} + \frac{j_\nu}{k}\left[1 - e^{-k(s_2-s_1)}\right] \quad (4.7)$$

The expression $k(s_2-s_1)$ appearing in the exponent is a dimensionless quantity known as the optical distance or thickness between the two points s_2 and s_1. As we shall see, it is often convenient to let optical depth rather than linear distance be the independent variable in radiation problems.

The two extreme cases of very large and very small optical depth are of particular interest. Equation (4.7) can be reduced to the following for these cases:

$$I_\nu(s_2) = I_\nu(s_1) + j_\nu(s_2 - s_1) \quad \text{(thin)}$$
$$I_\nu(s_2) = \frac{j_\nu}{k} \quad \text{(thick)} \quad (4.8)$$

In the optically thin case, we again simply have equation (3.2). In the optically thick case, the more distant parts of the source are not important because of the large absorption between them and the point in question. Only points within a few mean free paths give a significant contribution. In fact, an optically thick source is equivalent to a thin one whose size is one mean free path.

5. The Source Function

The ratio of the emission to the absorption coefficient is a very important quantity in radiation transfer theory; it is known as the source function:

$$S_\nu = \frac{j_\nu}{k} = \frac{\varepsilon_\nu}{\kappa} \quad (5.1)$$

One often works in terms of the source function instead of the emission coefficient. One reason for this is that the source function is usually far less sensitive to the properties of the medium than the emission coefficient. Another reason is that it fits in nicely with optical depth, when the latter is the independent variable. For example, the emission contribution of ds is $j_\nu ds = S_\nu d\tau$, where $d\tau = kds$ is the element of optical distance. As we shall see, for a range of circumstances wide enough to be useful in practice, the source function depends on temperature and frequency alone, and does not even depend on the composition of the medium.

Suppose that there are several different processes taking place, each with its own absorption and emission coefficients. The coefficients are additive, so $k = \Sigma k_i$, $j_\nu = \Sigma j_\nu(i)$. The total source function is then seen to be given by

$$S_\nu = \frac{j_\nu}{k} = \frac{1}{k}\sum j_\nu(i) = \frac{1}{k}\sum k_i S_\nu(i) \qquad (5.2)$$

The total source function is the sum of the individual ones, weighted by the absorption coefficients.

The following discussion is intended primarily for continuum processes. For bound-bound transitions, which are considered later, the emphasis is somewhat different, although much of what is stated here will also have application to the lines.

It is convenient to divide absorption and emission into scattering and non-scattering parts, indicated by sc and nc in what follows. A scattering process is one in which the photon essentially maintains its identity through the absorption and emission. Its direction and perhaps its frequency are changed, but there is a one to one correspondence between the photons before and after the scattering. For non-scattering, the photon is lost to the radiation field, at least temporarily, and there is only a statistical correlation between the photons being absorbed and those being emitted.

This distinction is important because the scattering part of the source function depends directly upon the radiation field, which is the main unknown in the problem. For non-scattering, however, the source function is only indirectly dependant upon the radiation field. The method of attacking a problem may vary considerably, depending on which types of processes dominate. From equation (5.2), we have

$$S_\nu = \frac{k(sc)}{k} S_\nu(sc) + \frac{k(ns)}{k} S_\nu(ns) \qquad (5.3)$$

Scattering processes will be considered first.

According to equation (2.1), the intensity of frequency ν' which is lost by scattering over the path ds is $I_{\nu'} k(sc) ds$. Then the energy lost from a pencil of radiation defined by the time interval dt, solid angle $d\omega'$, frequency interval $d\nu'$, and area $d\sigma$ is given by

$$dE_{\nu'}(sc) = I_{\nu'} k(sc) \cos\theta' \, ds \, d\sigma \, d\omega' \, d\nu' \, dt \qquad (5.4)$$

This energy is scattered into other directions and frequencies. Let the scattering be described by the phase function P, where $P(\Psi,\nu'\to\nu)d\nu d\omega/4\pi$ is the probability that, when a scattering does take place, the photon will be deflected through the angle Ψ into the solid angle $d\omega$, and the frequency will be changed from ν' to between ν and $\nu+d\nu$. P is normalized so that its integral over all directions and frequencies is 4π. Ψ is the scattering angle, the angle between the incident direction along $d\omega'$ and the scattered direction along $d\omega$. If (θ,ϕ) and (θ',ϕ') are the spherical angles of the scattered and incident directions, then

$$\cos \Psi = \cos \theta \cos \theta' + \sin \theta \sin \theta' \cos(\phi-\phi') \quad (5.5)$$

Notice that Ψ is symmetric in $d\omega$ and $d\omega'$, so that the scattering angle is the same if the roles of incident and scattered photons are interchanged.

We can now find from equation (5.4) and the phase function the amount of energy scattered into the new direction and frequency from the old pencil of radiation. Noting in equation (5.4) that $\cos \theta'$ $ds d\sigma = dV$, the volume element which is responsible for the scattering, we find

$$dE_\nu(\nu'\to\nu,\omega'\to\omega) = k(sc)I_{\nu'}P(\Psi,\nu'\to\nu) \, dVd\nu d\nu' \frac{d\omega d\omega'}{4\pi} dt \quad (5.6)$$

The energy scattered into the new frequency and solid angle intervals by dV from all incident radiation is obtained by integrating the above over all $d\nu'$ and $d\omega'$. But according to equation (3.1), this is $j_\nu(sc) \, dVd\nu d\omega dt$. Thus the scattered part of the emission coefficient is

$$j_\nu(sc) = \iint k(sc,\nu')I_{\nu'}(\theta',\phi')P(\Psi,\nu'\to\nu) \frac{d\nu' d\omega'}{4\pi} \quad (5.7)$$

The arguments of frequency and direction have been explicitly indicated in the above integrand. Equation (5.7) shows the direct way in which the scattering part of the emission coefficient, and therefore the source function, depend upon the radiation field.

Nothing further can be done with equation (5.7) unless the form of the phase function is known. One cannot solve a radiation transport problem unless the physics of the interaction is understood. If the scattering particles have relativistic velocities, or if the photons have energies which are comparable to the rest energy of the particles, then a significant frequency shift will take place, and the scattering is described by the Compton effect. Although these conditions do occur in a number of astrophysically interesting situations they do not occur in normal stellar atmospheres; they will not be considered here. The present concern is only with no frequency change. Frequency shifts due to the Doppler effect are of no importance for continuum transitions; as indicated in Chapter 4, they are very important for line transitions.

The phase function we are interested in is of the form $P(\Psi)\delta(\nu'-\nu)$, where δ is the Dirac δ function; it is zero everywhere except where the argument is zero, at which point it goes to infinity in such a way that an integral over it has the value of unity. The frequency integral in equation (5.7) then has the effect of changing all ν' values to ν. We find

$$j_\nu(sc) = k(sc,\nu)\int I_\nu(\theta',\phi')P(\Psi)\frac{d\omega'}{4\pi} \qquad (5.8)$$

From this we find the source function to be

$$S_\nu(sc) = \int I_\nu(\theta',\phi')P(\Psi)\frac{d\omega'}{4\pi} \qquad (5.9)$$

Note that both the emission coefficient and the source function will generally depend upon the direction.

The Rayleigh phase function, valid for ordinary Thomson scattering, is given by

$$P(\Psi) = \tfrac{3}{4}(1 + \cos^2\Psi) \qquad (5.10)$$

In most scattering problems it is felt that the direction dependence given by equation (5.10) is not important enough to justify the extra complications it brings into account. S. Chandrasekhar (Ap. J. 100, p 117, 1944) showed that, under certain circumstances, it does not have an important effect on the radiation field. Thus it is common practice to assume that the phase function has no direction dependence, $P = 1$,

$$S_\nu(sc) = \int I_\nu(\theta',\phi')\frac{d\omega'}{4\pi} = J_\nu \qquad (5.11)$$

The approximation of equation (5.11) is known as isotropic scattering.

We now turn our attention to the non-scattering part of the source function. Let two energy levels of an atom be designated 1 (lower) and 2 (upper). Transitions between these levels give rise to the absorption and emission of radiation and, therefore, to a contribution to the non-scattering part of the source function. There are three kinds of radiative transitions which can take place: spontaneous emission from 2 to 1; induced emission from 2 to 1; and absorption from 1 to 2. The rates at which these transitions take place are most conveniently expressed in terms of the Einstein coefficients. If B_{12}, B_{21}, and A_{21} are the Einstein coefficients of absorption, induced emission, and spontaneous emission, respectively, then

$$N(ab)d\nu d\omega = N_1 B_{12} I_\nu d\nu \frac{d\omega}{4\pi}$$
$$N(ie)d\nu d\omega = N_2 B_{21} I_\nu d\nu \frac{d\omega}{4\pi} \quad (5.12)$$
$$N(se)d\nu d\omega = N_2 A_{21} d\nu \frac{d\omega}{4\pi}$$

The left sides of these equations represent the number of transitions of the appropriate kind (ab = absorption; ie = induced emission; se = spontaneous emission) per unit volume and per unit time, in which the absorbed or emitted photon is in the frequency and direction ranges given by $d\nu$ and $d\omega$. N_1 and N_2 are the number of atoms per unit volume in the given energy levels. If E_{12} is the energy difference between these levels, then the frequency ν appearing in equations (5.12) is given by the constant E_{12}/h, where h is Planck's constant.

The Einstein coefficients given by equations (5.12) are atomic constants that depend on the atom and levels in question, but they do not depend on physical conditions. The coefficients are sometimes defined in a somewhat different way, such as in terms of energy density instead of intensity, and the effect is to change them by a multiplying constant.

If the levels belong to the continuum, then the abundances N_1 and N_2 are the number of electron-ion pairs per unit relative kinetic energy per unit volume, so they are proportional to the electron density as well as the ion density. If both levels are bound so the transition is a line, then a suitable broadening function as discussed in Chapter 4 must also be introduced to keep the Einstein coefficients independent of physical conditions.

Consider the volume element $dV = \cos\theta \, d\sigma ds$, where as usual ds makes the angle θ with the normal to $d\sigma$. The matter within dV modifies the radiation propagating through it by absorption and emission; the energy gained or lost equals the number of transitions times $h\nu$, so the net energy gain by our pencil of radiation due to dV follows from equations (5.12) to be

$$dE_\nu = \frac{h\nu}{4\pi}(N_2 A_{21} + N_2 B_{21} I_\nu - N_1 B_{12} I_\nu) \, dV d\nu d\omega dt \qquad (5.13)$$

Energy change is related to intensity change through equation (1.1). Evaluating dV as given above, we find for the equation of transfer

$$\frac{dI_\nu}{ds} = \frac{h\nu}{4\pi} N_2 A_{21} - \frac{h\nu}{4\pi}(N_1 B_{12} - N_2 B_{21}) I_\nu \qquad (5.14)$$

By comparing this with equation (4.1), we find the following expressions for the non-scattering contributions to the absorption and source function:

$$k(ns) = \frac{h\nu}{4\pi}(N_1 B_{12} - N_2 B_{21}) \qquad (5.15)$$

$$S_\nu(ns) = \frac{N_2 A_{21}}{N_1 B_{12} - N_2 B_{21}} = \frac{A_{21}/B_{21}}{\frac{N_1 B_{12}}{N_2 B_{21}} - 1} \qquad (5.16)$$

The non-scattering part of the source function depends only on the ratio of the populations of the energy levels, in addition to the values of the Einstein coefficients. To find the value of the source function, therefore, we must find what determines this population ratio.

Assume that conditions of thermodynamic equilibrium (TE) hold. In TE the radiation field is independent of position, so $dI_\nu/ds = 0$, and the intensity equals the source function. But in TE the intensity is given by the Planck function:

$$B_\nu(T) = \frac{2h\nu^3/c^2}{e^{h\nu/kT} - 1} \qquad (5.17)$$

Also, in TE the population ratio is given by the Boltzmann

excitation relation:

$$\frac{N_1}{N_2} = \frac{g_1}{g_2} e^{h\nu/kT} = \frac{g_1}{g_2} e^{E_{12}/kT} \tag{5.18}$$

Equations (5.16)-(5.18) lead to the following relations between the Einstein coefficients:

$$\frac{A_{21}}{B_{21}} = \frac{2h\nu^3}{c^2} \qquad \frac{B_{12}}{B_{21}} = \frac{g_2}{g_1} \tag{5.19}$$

The g's are the statistical weights of the levels, equal to the number of quantum mechanical states corresponding to the given energy.

The Einstein parameters are constants, independent of physical conditions; therefore, relations (5.19) are valid whether TE holds or not. If these are substituted into equation (5.16), the following is obtained for the source function:

$$S_\nu(ns) = \frac{2h\nu^3/c^2}{\frac{g_2 N_1}{g_1 N_2} - 1} \tag{5.20}$$

Although equation (5.18) is valid only in TE, one can define an excitation temperature T_{ex} by the relation

$$\frac{N_1}{N_2} = \frac{g_1}{g_2} e^{h\nu/kT_{ex}} \tag{5.21}$$

It follows that

$$S_\nu(ns) = B_\nu(T_{ex}) \tag{5.22}$$

The non-scattering part of the source function is the Planck function of the excitation temperature. In TE, the different measures of the temperature all agree, but in general this is not the case. If conditions are very far from TE, T_{ex} will have little semblance to other measures of temperature, and equation (5.22) has little physical content.

The kinetic temperature T_k describing the velocity distribution of the free particles is an important measure of temperature. Under most conditions of astrophysical interest elastic collisions are sufficiently numerous that the free particles follow closely a Maxwellian distribution, and T_k is a well defined quantity. If conditions are not too far from TE, T_{ex} for any set of levels will not differ much from T_k, and the non-scattering part of the source function becomes the Planck function of T_k. Whenever this is a sufficiently good approximation for all transitions of interest, the condition of local thermodynamic equilibrium (LTE) is said to hold.

In LTE the source function depends only on the kinetic temperature and frequency or the energy difference between the levels. As far as the source function is concerned, there is no difference between TE and LTE; however, the radiation field in LTE may be quite different from that in TE.

The analysis of a given situation to determine the validity of LTE is very complicated. The distribution of the atoms over the energy levels is determined by both collisional and radiative transitions. The collisional effects are fixed by the kinetic properties of the matter, so collisions tend to produce LTE. The radiative transitions are much more complicated. They would also drive the system toward LTE if the photons were incident uniformly from all directions and if they were in equilibrium with the local kinetic temperature. These conditions are met in the deep interior of a star, but the atmosphere of a star is the place where, by definition, these two conditions are not met. Thus it becomes necessary to solve the detailed equations of statistical equilibrium coupled with the equation of transfer inorder to test any given situation.

The consensus of opinion is that for a majority of stars, LTE in the continuum is a very good approximation. The main exceptions are extremely hot and extremely cool stars, very high luminosity stars, and stars with extended atmospheres. Review papers include B. E. J. Pagel, Proc. Roy. Soc. A. 306, 91, 1968 and D. Mihalas and R. G. Athay, Ann. Rev. Astron. Astroph. 11, 187, 1973. The line problem is even more complicated and will be considered in Chapter 4.

Unless otherwise stated, it will be here assumed that the non-scattering part of the continuum source function is in LTE. If the scattering part is described by isotropic scattering, then the total source function is

$$S_\nu = qB_\nu(T) + (1 - q)J_\nu \qquad (5.23)$$

where $q = k(ns)/k$ is the ratio of non-scattering to total absorption. T without subscript is kinetic temperature.

In TE the intensity is the Planck function, and both

scattering and non-scattering parts of the source function are given by the Planck function. This condition is closely approached deep inside a star.

For most stars k(ns) is much larger than k(sc), q is very close to unity, and the total continuum source function is that of LTE. In very hot stars, however, the large number of free electrons coming from the ionization of hydrogen causes electron scattering to be quite large, and both terms in equation (5.23) must be included.

6. Special Integrals for Plane Media

Most stellar atmospheres are extremely thin compared to the radii of the stars. Except when considering the extreme edge of the disk, therefore, one can ignore the curvature of the atmosphere when studying its radiation transport. This is a very great simplification in the geometry, as the equation of transfer can then be expressed in terms of the derivative of only one coordinate.

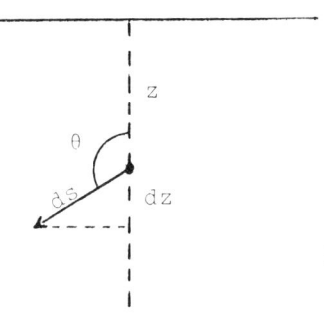

Figure 3. Geometry in a plane atmosphere

It is standard to measure distances and optical thicknesses positive into the star, but to measure angles from the outward direction (See Figure 3). This convention introduces a negative sign into the equation of transfer. Let z be the linear distance below the surface, and let θ be the angle from the outward direction. Then with ds directed along θ, we have

$$dz = -\cos \theta \, ds \qquad (6.1)$$

The equation of transfer (4.1) is then

$$\cos \theta \, \frac{dI_\nu}{dz} = kI_\nu - j_\nu = k(I_\nu - S_\nu) \qquad (6.2)$$

19

Replace linear depth z by optical depth τ, defined as

$$d\tau = k\,dz \qquad (6.3)$$

Note that τ, which is also zero at the surface of the star, is a measure of the perpendicular distance below the surface. In terms of optical depth, the equation of transfer (6.2) becomes

$$\mu \frac{dI_\nu}{d\tau} = I_\nu - S_\nu \qquad (6.4)$$

where $\mu = \cos\theta$.

The solution of equation (6.4) can be obtained by means of obvious modifications of equation (4.6). The result is most easily expressed if a distinction is made between upward and downward directions. This is caused by the different boundary conditions in the two directions, the atmosphere extending to infinity downward. The quantity $\mu = \cos\theta$, unless otherwise specified, will remain positive; the downward direction will be indicated by a minus sign. Thus $I(+\mu)$ and $I(-\mu)$ are the intensities into the two hemispheres. The solutions of the equation are then

$$I_\nu(\tau,+\mu) = \int_\tau^\infty S_\nu(\tau')\, e^{-(\tau'-\tau)/\mu}\, \frac{d\tau'}{\mu}$$

$$I_\nu(\tau,-\mu) = I_\nu(0,-\mu)\, e^{-\tau/\mu} + \int_0^\tau S_\nu(\tau')\, e^{-(\tau-\tau')/\mu}\, \frac{d\tau'}{\mu} \qquad (0\leq\mu\leq 1) \quad (6.5)$$

$I_\nu(0,-\mu)$ is the intensity incident on the top of the atmosphere. In most problems this is zero. In agreement with the usual situations, the source function has been assumed to be independent of direction.

The mean intensity, flux, and radiation pressure can also be found in terms of integrals of the source function over the atmosphere. If equations (6.5) are multiplied by the appropriate factors, the integrals of equations (1.3), (1.6), and (1.10) can be carried out. The results are given in terms of the exponential-integral function $E_n(x)$:

$$E_n(x) = \int_1^\infty e^{-xu}\, u^{-n}\, du = \int_0^1 e^{-x/v}\, v^{n-2}\, dv \qquad (6.6)$$

$$J_\nu(\tau) = \tfrac{1}{2}\int_0^\infty S_\nu(\tau')E_1(|\tau' - \tau|)\,d\tau' \tag{6.7}$$

$$F_\nu(\tau) = 2\pi\int_\tau^\infty S_\nu(\tau')E_2(\tau' - \tau)\,d\tau' - 2\pi\int_0^\tau S_\nu(\tau')E_2(\tau - \tau')\,d\tau' \tag{6.8}$$

$$P_{r_\nu}(\tau) = \frac{2\pi}{c}\int_0^\infty S_\nu(\tau')E_3(|\tau' - \tau|)\,d\tau' \tag{6.9}$$

The exponential-integral functions are somewhat similar to the ordinary exponential function. $E_n(0) = 1/(n-1)$, and $E_n(x) \to e^{-x}/x$ as $x \to 0$. It is seen that E_1, E_2, and E_3 take the part of the exponential attenuation for mean intensity, flux, and radiation pressure, respectively. In the above equations the radiation incident on the top of the atmosphere has been ignored.

An important energy conservation relation can be obtained from equation (1.4) by taking its divergence. In the integrand of that equation only the intensity is a function of the position coordinates, so only the intensity is operated upon:

$$\nabla \cdot \vec{F}_\nu = \int \hat{s} \cdot \nabla I_\nu\, d\omega \tag{6.10}$$

But the integrand is the component of the gradient of intensity along $\hat{s}$, which is dI_ν/ds. From the equation of transfer this is seen to be $k(S_\nu - I_\nu)$. The integral over solid angle gives

$$\nabla \cdot \vec{F}_\nu = 4\pi k(S_\nu - J_\nu) \tag{6.11}$$

In a plane medium, the vector flux is directed up out of the star, so the divergence becomes $-d/dz$. Changing to optical depth, we find

$$\frac{dF_\nu}{d\tau} = 4\pi(J_\nu - S_\nu) \tag{6.12}$$

Equations (6.11) and (6.12) simply state that the flux at any frequency results from the excess of emissions over absorptions at that frequency.

If the equation of transfer (6.4) is multiplied by μ and integrated over solid angle, one finds

$$\frac{dP_{r\nu}}{d\tau} = \frac{F_\nu}{c} \qquad (6.13)$$

We are again making the assumption that the source function does not depend upon direction. If another derivative of this equation with respect to optical depth is taken and equation (6.12) is used, we find

$$\frac{d^2 P_{r\nu}}{d\tau^2} = \frac{4\pi}{c}(J_\nu - S_\nu) \qquad (6.14)$$

These relations will be referred to in later sections.

Chapter 2: The Gray Atmosphere

7. Introduction

In the present chapter the concern will be with stellar atmospheres which have the following characteristics: 1. static 2. semi-infinite; 3. plane; 4. no radiation incident on top; 5. LTE; 6. energy carried only by radiation; 7. the absorption coefficient independent of frequency. A gray atmosphere is usually defined as one which satisfies condition 7., but for convenience, I shall use the term gray in this chapter to include all seven conditions.

The first five conditions mentioned above need no elaboration at present beyond the discussion in Chapter 1. Condition 6 is generally called radiative equilibrium. It means that radiative energy is conserved, so that whatever radiative energy is put in at the bottom of the atmosphere must come out the top. For a plane atmosphere, the flux carried by radiation is a constant with depth. Radiative equilibrium is a good approximation for most atmospheres, as the nuclear energy sources are far removed from the atmosphere, and radiation is usually the dominant form of energy transfer in the atmosphere. There are some cases, however, when convection may carry a significant fraction of the energy through the atmosphere. Then the exchange of energy between radiation and convection must be taken into account. The problem of convection will be discussed in the next chapter.

The assumption that the absorption coefficient is independent of frequency is highly artificial. Historically this assumption was often made because it represented a tremendous simplification and because it offered a large amount of physical insight on the more complicated problems of real atmospheres. Besides, the correct variation of absorption with frequency was not well known. It is now known that the absorption throughout the most important frequency regions of moderate temperature stars is dominated by the negative ion of hydrogen, and H⁻ absorption is very nearly independent of frequency. Thus the gray atmosphere turns out to be a much better representation of real stars than one could have expected.

With LTE the equation of transfer (6.4) becomes

$$\mu \frac{dI_\nu}{d\tau} = I_\nu - B_\nu \qquad (7.1)$$

where the optical depth is now independent of frequency. This relation can now be integrated over frequency, something which will not produce a meaningful result in the non-gray case. The result is

$$\mu \frac{dI}{d\tau} = I - B \qquad (7.2)$$

The integrated Planck function is given by

$$B = \frac{\sigma}{\pi} T^4 \qquad (7.3)$$

where σ is the Stefan-Boltzmann radiation constant (5.67×10^{-5} erg cm^{-2} K^{-4}).

In the gray case the flux integral (6.12) can be integrated over frequency. For a plane atmosphere in radiative equilibrium the radiative flux is constant with depth, $dF/d\tau = 0$. Note that this is true for the integrated flux, but not the monochromatic flux. It follows that

$$B = J \qquad (7.4)$$

Again, note that B_ν does not equal J_ν. Equations (6.7) and (6.8) integrated over frequency give the following for the mean intensity and flux:

$$J(\tau) = B(\tau) = \frac{1}{2} \int_0^\infty B(\tau') E_1(|\tau' - \tau|) \, d\tau' \qquad (7.5)$$

$F(\tau)$ = constant

$$= 2\pi \int_\tau^\infty B(\tau')E_2(\tau' - \tau)\, d\tau' - 2\pi \int_0^\tau B(\tau')E_2(\tau - \tau')\, d\tau' \quad (7.6)$$

The effective temperature T_e of a star is the temperature of a black body which radiates energy at the same rate per unit area as the star. In terms of the surface flux, this is

$$F = \sigma T_e^4 \quad (7.7)$$

Equations (7.5) and (7.6) suffice to uniquely determine $B(\tau)$ for a given value of F or T_e. But B is a known function of the temperature, so the temperature-optical depth relation is also determined. It follows that the monochromatic Planck function B_ν is fixed as a function of depth and, finally, equations (6.5) determine the monochromatic intensity for any depth and direction. In terms of the single input parameter F (or T_e), the entire radiation field of a gray atmosphere is fixed.

It may seem surprising that values of the absorption coefficient k are not needed in order to find the radiation field. This is essentially due to two factors: 1. optical depth rather than linear depth is used as the independent variable; and 2. the absorption is known to be the same for all frequencies. If one wants the solution in terms of the linear depth z, then numerical values of k are needed and the problem becomes much more difficult.

If k is large, most of the energy escaping from the atmosphere will come from regions near the surface in linear distance; if k is small, regions at a much greater linear distance from the surface will be important. In both cases the regions near optical depth unity make the most contribution to the emergent radiation. If k changes with frequency, a region may be at optical depth unity at one frequency and much deeper or shallower at another frequency. In order to determine the energy balance over all frequencies, it is necessary to tie together these different regions through the absorption coefficient, i.e., k must be known. But when k is not a function of frequency, optical depth unity at one frequency is the same for all frequencies, and further information about k is not needed.

The exact solution for the gray temperature distribution was first given by C. Mark, Phys. Rev. 72, 558, 1947. In the next section a very useful approximation method will be applied to the gray problem, and this will be followed by a more accurate method of approach. In both cases the methods have many applications beyond the simple gray problem considered here.

8. The Eddington Approximation

Let us for the moment set aside the gray atmosphere and consider the equation of transfer for a general plane semi-infinite atmosphere:

$$I_\nu(\tau,\mu) = S_\nu(\tau) + \mu\frac{dI_\nu}{d\tau} \qquad (8.1)$$

As we go to very great depths in the star, the effects of the boundary will diminish and the intensity will become nearly isotropic, i.e., nearly independent of direction. Now the derivative term on the right side of equation (8.1) depends directly upon direction through $\mu = \cos\theta$. In the near isotropy of the deep levels, this term must be very small; therefore, $I_\nu \to S_\nu$ for very large optical depths. Under these conditions, we can treat the derivative term as a perturbation, and in it substitute the source function for the intensity:

$$I_\nu(\tau,\mu) \simeq S_\nu(\tau) + \mu\frac{dS_\nu}{d\tau} \qquad (8.2)$$

If we substitute this approximate form of the intensity into equations (1.3), (1.6), and (1.10), we obtain the following:

$$J_\nu \simeq S_\nu$$

$$F_\nu \simeq \frac{4\pi}{3}\frac{dS_\nu}{d\tau} \simeq \frac{4\pi}{3}\frac{dJ_\nu}{d\tau} \qquad (8.3)$$

$$P_{r\nu} \simeq \frac{4\pi}{3c}S_\nu \simeq \frac{4\pi}{3c}J_\nu$$

We see from the above that the second derivative term in equation (6.14) is also very small at great depths; we can substitute mean intensity for radiation pressure as indicated by the third relation of (8.3):

$$\frac{d^2 J_\nu}{d\tau^2} \simeq 3(J_\nu - S_\nu) \qquad (8.4)$$

This relation is very accurate at large depths in a star. The Eddington approximation consists of the assumption that it is valid at all depths, even up to the surface. It should be noted that nothing has been said about radiative equilibrium, LTE, or the frequency dependence of the absorption coefficient. The Eddington approximation is quite independent of the assumption of grayness.

Boundary conditions appropriate to the Eddington approximation may be found in a number of ways, all about equally good but not all equivalent. A useful approximate condition involves the surface flux:

$$F_\nu(0) = 2\pi \int_0^1 I_\nu(0,\mu)\mu d\mu \qquad (8.5)$$

If we replace μ in the integrand by its average value of 0.5, we obtain

$$F_\nu(0) \simeq \pi \int_0^1 I_\nu(0,\mu) d\mu = 2\pi J_\nu(0) \qquad (8.6)$$

Another boundary condition follows from equation (8.6) and the second relation of (8.3) applied to the surface:

$$J_\nu(0) \simeq \frac{1}{2\pi} F_\nu(0) \simeq \frac{2}{3} \left.\frac{dJ_\nu}{d\tau}\right)_0 \qquad (8.7)$$

The Eddington approximation will now be applied to the gray atmosphere. As optical depth does not depend upon frequency, equation (8.4) can be integrated over frequency. With equation (7.4), we find

$$\frac{d^2 J}{d\tau^2} = 0 \qquad (8.8)$$

The solution is

$$J = B = a\tau + b \tag{8.9}$$

where a and b are constants. Condition (8.7) requires a = 3b/2, while relation (8.6) leads to b = F/2π. The depth dependence of J and B is then found to be

$$J = B = \frac{3F}{4\pi}\left(\tau + \frac{2}{3}\right) \tag{8.10}$$

This can also be expressed in terms of temperatures by using equations (7.3) and (7.7):

$$T^4 = \frac{3}{4} T_e^4 \left(\tau + \frac{2}{3}\right) \tag{8.11}$$

The surface temperature is seen to be $(0.5)^{1/4} = 0.841$ times the effective temperature.

The temperature is known at each depth, so equations (6.5) can be used to find the monochromatic intensity for any depth and direction. Thus the entire radiation problem has been solved (in a certain approximation) in terms of the single parameter F (or equivalently T_e). Equation (8.11) shows how the temperature distribution scales for atmospheres of different effective temperatures.

The integrated surface intensity is

$$I(0,\mu) = \frac{3F}{4\pi}\left(\mu + \frac{2}{3}\right) \tag{8.12}$$

A comparison of equations (8.10) and (8.12) shows that the surface intensity is equal to the source function evaluated at an optical depth equal to cos θ, that is, at an optical depth measured along the line of sight of unity. This is a very useful relation when high accuracy is not needed. The above equation (8.12) also gives the limb darkening in the Eddington approximation, the amount by which intensity varies at different points on the disk of the star.

The accuracy of the Eddington approximation is illustrated in Table 1, where the exact gray temperature distribution is compared with that of the Eddington approximation. As expected, the largest errors occur at and near the surface, and they become small at great depths.

Table 1.

Temperatures in Exact Gray and Edd. Approx. Atmospheres

τ	T(gray)/T_e	T(gray)/T(Edd)
0.0	0.8112	1.036
0.05	0.8391	1.020
0.1	0.8595	1.013
0.2	0.8934	1.005
0.5	0.9699	0.997
1	1.0625	0.995
2	1.1938	0.996
5	1.4386	0.998

It is worth noting that the upper layers in which the Eddington approximation has the greatest errors are not very important in the emission of radiation, as the main contribution to the emergent radiation comes from layers about one mean free path into the star. In the monochromatic flux, for example, there is less than one percent difference between the exact gray and the Eddington approximation solutions. This would not be the case for intensities coming from near the limb of a star, as here the shallow layers are important.

9. The Method of Discrete Ordinates

The discussion in this section is based largely on the development by S. Chandrasekhar in his book Radiative Transfer. Equations (7.2) and (7.4) show that for a gray atmosphere,

$$\mu \frac{dI}{d\tau} = I - \frac{1}{2} \int_{-1}^{+1} I(\tau, \mu') \, d\mu' \qquad (9.1)$$

Integration over the azimuth angle φ has already been carried out, as there is azimuthal symmetry in a plane atmosphere. The boundary conditions are that there is no radiation incident on the surface: $I(0,-\mu) = 0$; and that no radiation can survive from infinite optical depths: $I(\tau)e^{-\tau} \to 0$ as $\tau \to \infty$.

Chandrasekhar's method of solving the integral-differential equation (9.1) is to replace the integral by a summation. A technique is chosen so that the maximum accuracy is obtained for a given number of terms in the sum. The Gaussian quadrature method suggests itself, and a brief digression on this method at this stage will be useful.

The m-point quadrature formula for use with equation (9.1) is of the form

$$\int_{-1}^{+1} f(\mu)d\mu \simeq \sum_{i=1}^{m} a_i f(\mu_i) \qquad (9.2)$$

The Gaussian method consists of taking for the points of division μ_i the zeros of the mth order Legendre polynomial:

$$P_m(\mu_i) = 0, \quad i = 1, 2, \ldots, m \qquad (9.3)$$

The weights a_i are given by

$$a_i = \frac{1}{P_m'(\mu_i)} \int_{-1}^{+1} \frac{P_m(\mu)}{\mu - \mu_i} d\mu \qquad (9.4)$$

The prime here indicates the derivative. For these choices of the a_i and μ_i the integral is given exactly by the sum if the function $f(\mu)$ is a polynomial of degree $(2m-1)$ or less. For more general functions the Gaussian sum gives a far better representation of the integral, for a given number of terms, than the usual techniques which have the μ_i equally spaced. It is the most accurate method for given m for functions which are well represented by polynomials.

It is convenient to restrict m to even integers, $m = 2n$. In this case the Legendre polynomial $P_{2n}(\mu)$ contains only even powers of μ, and the roots occur in pairs $\pm\mu_i$; thus the roots are symmetric between the inward and the outward hemispheres. From equation (9.4) one can show that the weights associated with the two roots $\pm\mu_i$ are equal; it then becomes convenient to designate the weights and points by (a_i, μ_i), where

$$a_i = a_{-i} \qquad \mu_i = -\mu_{-i} \qquad i = \pm 1, \pm 2, \ldots, \pm n \qquad (9.5)$$

The first three sets of Gaussian weights and points are given in Table 2.

Table 2

Gaussian Weights and Points

$n = 1$	$\mu_{\pm 1} = \pm 0.577350$	$a_{\pm 1} = 1.000000$
$n = 2$	$\mu_{\pm 1} = \pm 0.339981$	$a_{\pm 1} = 0.652145$
	$\mu_{\pm 2} = \pm 0.861136$	$a_{\pm 2} = 0.347855$
$n = 3$	$\mu_{\pm 1} = \pm 0.238619$	$a_{\pm 1} = 0.467914$
	$\mu_{\pm 2} = \pm 0.661209$	$a_{\pm 2} = 0.360762$
	$\mu_{\pm 3} = \pm 0.932470$	$a_{\pm 3} = 0.171324$

Since equation (9.2) is exact for $f(\mu)$ a polynomial of degree $(4n-1)$ or less, an important property of the weights and points is

$$\int_{-1}^{+1} \mu^p d\mu = \sum_{i=\pm 1}^{\pm n} a_i \mu_i^p = \frac{2}{p+1} \quad \text{(p even)}$$
$$= 0 \quad \text{(p odd)} \qquad p \leq (4n-1) \qquad (9.6)$$

The application of this method to the gray problem begins with the replacement of the integral in equation (9.1) by a Gaussian sum of order n:

$$\mu \frac{dI}{d\tau} = I - \frac{1}{2} \sum_{j=\pm 1}^{\pm n} a_j I_j \qquad (9.7)$$

I_j represents $I(\tau,\mu_j)$. To the extent that the above sum is an accurate representation of the integral, equation (9.7) is valid for all depths and directions.

Equation (9.7) is a linear, homogeneous differential equation, so a solution will be sought of the form

$$I(\tau,\mu) = g(\mu)e^{-k\tau} \qquad (9.8)$$

where k is a constant. If this is substituted into equation (9.7), the condition on $g(\mu)$ is found to be

$$g(\mu) = \frac{\frac{1}{2}\sum_{i=\pm 1}^{\pm n} a_i g_i}{1 + k\mu} = \frac{\text{constant}}{1 + k\mu} \qquad (9.9)$$

If this relation is evaluated at $\mu = \mu_j$, each term multiplied by a_j and summed over all j values, we obtain the following condition on k:

$$\sum_{j=\pm 1}^{\pm n} \frac{a_j}{1 + k\mu_j} = 2 \qquad (9.10)$$

By using the conditions (9.5), we can transform the above to

$$\sum_{j=1}^{n} \frac{a_j}{1 - k^2\mu_j^2} = 1 \qquad (9.11)$$

The sum in equation (9.11) is taken only over positive values of j.

Equation (9.11) is equivalent to a polynomial equation of degree n in k^2, and there are thus 2n roots for k which occur in ± pairs. One of the solutions is $k^2 = 0$, however, as can be seen from equation (9.6) for p = 0. Thus there are only (2n-1) separate solutions of the form (9.8). The (2n-2) non-zero values of k will be designated as k_m, m = ±1,...,±(n-1), with $k_m = -k_{-m}$. The k=0 solution will be denoted Q. The 2nth independent solution of equation (9.7) is seen by inspection to be the quantity $(\tau+\mu)$. The general solution of equation

(9.7) is a linear combination of the 2n particular solutions. It can be shown, however, that the particular solutions which correspond to negative values of k violate the boundary condition at large depths, as $k_m^2 > 1$ for all m. Thus the coefficients of these (n-1) terms must be set to zero, and we have

$$I(\tau,\mu) = A\left[\sum_{m=1}^{n-1} \frac{L_m e^{-k_m \tau}}{1 + k_m \mu} + Q + \tau + \mu\right] \quad (9.12)$$

A, Q, and L_m, m = 1,...,n-1, are the remaining integration constants.

The upper boundary condition is that there be no radiation incident on the top of the atmosphere:

$$\sum_{m=1}^{n-1} \frac{L_m}{1 - k_m \mu} + Q - \mu = 0 \quad 0 < \mu \leq 1 \quad (9.13)$$

It can be seen that this condition cannot be satisfied for all inward directions. The best that can be done is to use the n constants L_m and Q to satisfy it for n values of μ. There is an obvious advantage for choosing as these n directions the points of division of the Gaussian sum μ_i, i = 1,...,n; in this way the sum in equation (9.7) will more accurately represent the integral. Instead of equation (9.13), therefore, we have the L_m and Q determined by

$$\sum_{m=1}^{n-1} \frac{L_m}{1 - k_m \mu_i} + Q - \mu_i = 0 \quad i = 1,...,n \quad (9.14)$$

The final constant to be determined is A, and it is related to the flux.

$$F(\tau) = 2\pi \int_{-1}^{+1} I(\tau,\mu)\mu d\mu \simeq 2\pi \sum_{i=\pm 1}^{\pm n} a_i \mu_i I(\tau,\mu_i) \quad (9.15)$$

When equation (9.12) and condition (9.6) are used, we find

$$F(\tau) = 2\pi A \left[\sum_{m=1}^{n-1} L_m e^{-k_m \tau} \left(\sum_{i=\pm 1}^{\pm n} \frac{a_i \mu_i}{1 + k_m \mu_i} \right) + \frac{2}{3} \right] \qquad (9.16)$$

But from equations (9.6) and (9.10),

$$\sum_{i=\pm 1}^{\pm n} \frac{a_i \mu_i}{1 + k_m \mu_i} = \frac{1}{k_m} \sum_{i=\pm 1}^{\pm n} a_i \left(1 - \frac{1}{1 + k_m \mu_i} \right) = 0 \qquad (9.17)$$

The multiplying constant A is found to be $3F/4\pi$. The intensity is then

$$I(\tau,\mu) = \frac{3F}{4\pi} \left[\sum_{m=1}^{n-1} \frac{L_m e^{-k_m \tau}}{1 + k_m \mu} + Q + \tau + \mu \right] \qquad (9.18)$$

It should again be emphasized that equation (9.18) can be applied to all upward directions, but only those downward directions which are along the points of division satisfy the boundary conditions. In fact equation (9.18) is seen to be singular for those directions given by $\mu = -1/k_m$, but these are not the directions of the $-\mu_i$.

The temperature distribution in the atmosphere can be found by evaluating the mean intensity:

$$J(\tau) = B(\tau) = \frac{1}{2} \int_{-1}^{+1} I(\tau,\mu) \, d\mu \simeq \frac{1}{2} \sum_{i=\pm 1}^{\pm n} a_i I(\tau,\mu_i) \qquad (9.19)$$

If this is evaluated from equation (9.18), it is found that

$$J(\tau) = B(\tau) = \frac{3F}{4\pi} \left[\sum_{m=1}^{n-1} L_m e^{-k_m \tau} + \tau + Q \right] \qquad (9.20)$$

If this result is compared with equation (8.10), we see that the Eddington approximation to the gray atmosphere corresponds to $L_m = 0$, $Q = 2/3$. The integration constants for the first three approximations are listed in Table 3.

Table 3

Integration Constants for First Three Approximations

n = 1   Q = 0.577350

n = 2   Q = 0.694025 k_1 = 1.972027 L_1 = -0.116675

n = 3   Q = 0.703899 k_1 = 1.225211 L_1 = -0.025304
                       K_2 = 3.202945 L_2 = -0.101245

Equation (9.20) is commonly written in the form

$$B(\tau) = \frac{3F}{4\pi}\left[\tau + q(\tau)\right] \quad (9.21)$$

where $q(\tau)$ is a slowly varying function of τ. Exact values of q are given in Table 4.

Table 4

Exact Values of $q(\tau)$

τ	$q(\tau)$	τ	$q(\tau)$
0.00	0.57735	0.5	0.68029
.05	.61076	1.0	.69854
.10	.62792	1.5	.70513
.20	.64955	2.0	.70792
.30	.66336	∞	.71045

We can obtain an expression for the intensity which is valid for all directions; we use equation (9.20) as the source function and evaluate the integrals in equations (6.5). The result agrees exactly with equation (9.18) for upward directions; however, for downward directions we find a new expression:

$$I(\tau,-\mu) = \frac{3F}{4\pi} \sum_{m=1}^{n-1} \frac{L_m}{1 - k_m\mu} (e^{-k_m\tau} - e^{-\tau/\mu}) +$$

$$+ \tau + (Q - \mu)(1 - e^{-\tau/\mu}) \qquad (9.22)$$

The singularities at $\mu = -1/k_m$ have been removed in this equation; also, it is seen that equation (9.22) exactly obeys the boundary condition that $I(0,-\mu) = 0$ for all inward directions.

Although the quantities considered so far are integrated over all frequencies, monochromatic quantities can also be obtained. Equation (9.20) fixes the temperature distribution, and so B_ν is known at all depths for any desired frequency. The desired monochromatic quantity can then be found by using the appropriate relation from Section 6. The integrated flux F in this problem is constant with depth, so the increase of temperature into a star requires $F_\nu(\tau)$ have its maximum shift to higher frequencies at greater depths, the total area under the curve remaining constant.

Figure 4 compares the emitted flux from a gray atmospere with that observed from the Sun. The observations are taken from C. W. Allen, Astrophysical Quantities, 3rd ed., Athlone Press, 1973. We see from this that the Sun is indeed reasonably well represented by a gray atmosphere, the maximum error being of the order of 10%. It is worth noting that the differences between the exact gray flux and that in the Eddington approximation are much less that one percent at most frequencies, much too small to be detected in this figure.

10. Isotropic Scattering

For pure isotropic scattering without non-scattering, the source function is

$$S_\nu = J_\nu \qquad (10.1)$$

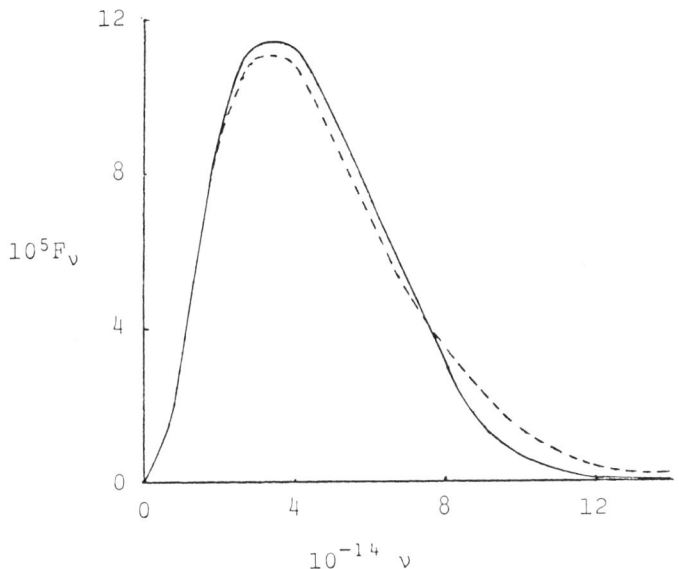

Figure 4. The flux emitted by the Sun according to observations (solid curve) and according to the gray model (dotted curve). The units are cgs.

The equation of transfer in a plane atmosphere is then

$$\mu \frac{dI_\nu}{d\tau} = I_\nu - J_\nu \qquad (10.2)$$

This is identical in form with equations (7.2) and (7.4) for the gray problem, the only difference being that equation (10.2) is in terms of monochromatic quantities. The boundary conditions are the same, so the functions $I(\tau,\mu)$, $J(\tau)$, etc., which are solutions of the gray problem must be mathematically identical to $I_\nu(\tau,\mu)$, $J_\nu(\tau)$, etc., which are solutions of the isotropic scattering problem. The relations of Section 9 can thus be applied here, and the scattering problem has already been solved. The fact that optical depth may depend upon the frequency in the scattering problem does not change the mathematical form of the solution.

In the gray case, the source function was equal to the mean intensity J only if we have radiative equilibrium, i.e., only if F is constant with depth. It follows that we expect equation (10.2) to also require F_ν to be constant with depth, and a term by term integration of the relation over solid angle shows that this is indeed true. We can see why this must be the case, for the assumed scattering does not change the frequency of a photon, and scattering is all that can happen to a photon. Thus there is no coupling between different frequencies or with other forms of energy. Whatever radiative flux is put into the bottom of the atmosphere must come out the top at the same frequency.

The gray atmosphere in radiative equilibrium and the isotropic scattering atmosphere are quite different physically though they have the same mathematical solution. In the gray case, energy absorbed at one frequency can be emitted at another, and the temperature is distributed with depth in such a way that the flux carried by all frequencies is constant. There is only one independent parameter, the flux; the solution easily scales from one flux to another.

In the scattering case the temperature is irrelevant. The energy absorbed at one frequency is emitted at the same one, and the source function must find a distribution with depth that has a constant flux at the same frequency. There is one independent parameter for each frequency, F_ν.

Chapter 3: The Non-Gray Atmosphere

11. The Model Atmosphere

When the absorption coefficient depends on frequency, the procedures of the last chapter are not successful in determining the radiation field. As pointed out in Section 7, it is necessary to relate the absorptions at different frequencies, and this usually means that we must evaluate the coefficient itself. Thus the equations which describe the physical state of the atmosphere must be solved along with the equation of transfer, and the degree of complexity is greatly increased. A listing of the temperature, pressure, density, and other important physical quantities as functions of depth is known as a model atmosphere. Usually it is necessary to construct a model atmosphere in order to solve for the radiation field.

The usual procedure for trying to solve the stellar atmosphere problem is to use an iteration technique. As we shall see, far from the boundary the solution simplifies to the same form as the gray problem; however, the atmosphere is the region where, by definition, the boundary is near enough to make itself known. One first constructs a trial model atmosphere, and then tests it for consistency with its radiation field. The calculated departures from radiative equilibrium then serve as clues for improving the trial model. In this section, we are concerned with determining the trial model.

For most stellar atmospheres, hydrostatic equilibrium is a very good approximation. This means that there are no large unbalanced forces on the material, so the pressures or, more accurately the pressure gradients, push upward and balance the force of gravity pushing downward. If P is the total pressure, z the linear distance into the star, ρ the mass density, and g the acceleration due to gravity, then hydrostatic equilibrium requires that

$$dP = g\rho dz \qquad (11.1)$$

If the star is rapidly rotating, g must include the effects of centrifugal acceleration and will vary from point to point on the star. Otherwise g includes only the effects of gravity and equals its surface value for a thin atmosphere:

$$g = \frac{GM}{R^2} \qquad (11.2)$$

M and R are the mass and radius of the star, and G is the gravitational constant. For the relatively few stars having thick atmospheres, the variation of g with depth should be taken into account; however, for these stars curvature of the atmosphere is also important, and the plane approximation should also be given up.

The pressure P includes contributions from both the gas pressure P_g and the radiation pressure P_r. If mass motions are present, the hydrodynamic pressure $\rho v^2/2$ may make a contribution, although it is usually very small. The magnetic pressure $H^2/8\pi$ can also be important in regions of large magnetic fields.

If only gas and radiation pressure are important, we have

$$\frac{dP_g}{dz} + \frac{dP_r}{dz} = g\rho \qquad (11.3)$$

Since radiation pressure depends directly upon the radiation field, it complicates the situation. Radiation pressure affects matter in that, when a photon is absorbed, its momentum is transferred to the atom which absorbs it. Referring back to equation (1.1) and including a frequency dependence, we see that the momentum transferred to the material within the volume element $dV = d\sigma \cos \theta\, ds$ by a given pencil of radiation is

$$d\vec{p}_\nu(\omega) = \frac{\hat{s}}{c} dE_\nu(abs) = \frac{\hat{s}}{c} kI_\nu dVd\omega d\nu dt \qquad (11.4)$$

$$d\vec{p} = \frac{dVdt}{c} \int\int k\hat{s}I_\nu d\omega d\nu = \frac{dVdt}{c} \int k\vec{F}_\nu d\nu \qquad (11.5)$$

Here $\hat{s}$ is a unit vector along the path of propagation. Now let dV consist of a horizontal area dA and a vertical thickness dz; then we find the radiation pressure by dividing the above by dAdt:

$$dP_r = \frac{dz}{c} \int_0^\infty kF_\nu d\nu \qquad (11.6)$$

We finally get for equation (11.3)

$$\frac{dP_g}{dz} = \rho(g - \frac{1}{c}\int_0^\infty \kappa F_\nu d\nu) \qquad (11.7)$$

Here we have changed from the volume to the mass absorption coefficient. It is usually more convenient to use the mass coefficient in stellar atmospheres. We see that radiation pressure has the effect of reducing the value of the effective surface gravity of the star.

It is sometimes possible to make a simplification of the above. Define a mean absorption coefficient or opacity as

$$\kappa_o = \frac{1}{F}\int_0^\infty \kappa F_\nu d\nu \qquad (11.8)$$

Equation (11.7) then becomes

$$\frac{dP_g}{dz} = \rho(g - \frac{\kappa_o F}{c}) \qquad (11.9)$$

If we introduce the mean optical depth $d\tau_o = \kappa_o \rho dz$, and if we replace the flux by the effective temperature, we find

$$\frac{dP_g}{d\tau_o} = \frac{g}{\kappa_o} - \frac{\sigma T_e^4}{c} \qquad (11.10)$$

It would appear that we have not gained anything by introducing the opacity, as it depends on the radiation field. We will see in the next section, however, that it is possible to find an opacity which approximately satisfies equation (11.8) without prior knowledge of the flux. If this approximation is sufficient, then equation (11.10) can be used directly. If not, then an iterative procedure must be used with the radiation pressure term.

Radiation pressure may be neglected for all except the hottest stars, as can be seen from the following rough calculation. Equation (1.10) indicates that in LTE and in layers that are not too near the surface, the radiation pressure is approximately

$$P_r \simeq \frac{4\pi}{3c} B(T_e) = \frac{4\sigma}{3c} T_e^4 \qquad (11.11)$$

If P_r is small, equation (11.10) shows that the typical value of the gas pressure in an atmosphere is

$$P_g = g \int_0^1 \frac{d\tau_o}{\kappa_o} \simeq \frac{g}{\kappa_o} \qquad (11.12)$$

In this, the opacity is a typical value for the atmosphere as a whole. This is of the order of 1 cm²/g, which leads to a gas pressure which is numerically about the same in cgs units as the value of the surface gravity. From this and equation (11.11), we find

$$\frac{P_r}{P_g} \simeq 10^{-15} \frac{T_e^4}{g} \qquad (11.13)$$

Again, cgs units are used. Most stars have surface gravities which are about 10^4 cgs; the above suggests that radiation pressure is less than a 10% effect if the effective temperature is under 30,000 K. More accurate calculations substantially agree with this conclusion. From here on, the radiation pressure term will be considered negligible.

With $P_g = P$, the total pressure, we need not be concerned with equation (11.8). Instead let κ_o be an opacity which is related to the absorption coefficient in a way that is yet to be specified. Then the hydrostatic equilibrium condition reduces to

$$\frac{dP}{d\tau_o} = \frac{g}{\kappa_o} \qquad (11.14)$$

Now let N_{ijn} be the number of particles per unit mass of the stellar material of element i, in ionization stage j, and in excitation level n. Also let a_{ijn} be the atomic absorption coefficient in cross sectional area for a particle of type (i,j,n). This cross section will depend on the frequency of the radiation. Then in analogy with equation (2.6), the total mass absorption coefficient at a given frequency is

$$\kappa = \sum_{ijn} N_{ijn} a_{ijn} \qquad (11.15)$$

The cross sections a_{ijn} depend on the properties of the different levels, as well as on the frequency. The number of particles per unit mass N_{ijn} depend on the chemical abundances and on how the atoms are distributed over the different energy states. In LTE the latter is determined by the temperature and the electron pressure, through the excitation and the ionization equations. Under these conditions, we can write the above equation (11.15) in the schematic form

$$\kappa = \kappa(\nu, z_i, T, P_e) \qquad (11.16)$$

This relation simply indicates that we can calculate the value of the absorption coefficient, at least in principle, if we know the values of the quantities in parentheses. z_i is the abundance of the ith element. The opacity κ_o is related in some way to the absorption coefficient, so a similar relation can be expressed for it.

Let x_i be the average number of free electrons which are supplied by nucleus i. Then the electron pressure is related to the total pressure by the relation

$$\frac{P_e}{P} = \frac{\sum_i z_i x_i}{\sum_i z_i (1 + x_i)} \qquad (11.17)$$

where z_i is the abundance by number of atoms of element i. The x_i's can be found from the ionization relations as functions of temperature and electron pressure. Thus equation (11.17) can also be expressed in the schematic form

$$P_e = P_e(z_i, P, T) \qquad (11.18)$$

Through this relation the independent variable in equation (11.16) can be changed from P_e to P:

$$\kappa = \kappa(\nu, z_i, T, P) \qquad (11.19)$$

Let us assume for the moment that the relation between temperature and optical depth τ_o is known. Then if the abundances are also known, the absorption becomes a known function of optical depth and pressure through relation (11.19). But pressure and optical depth are the integration variables in the hydrostatic equilibrium equation (11.14), so this equation can be integrated. This provides T, P, and κ_o as functions of τ_o. (It is assumed that we have decided how we wish to make the opacity κ_o depend upon the absorption coefficient κ.) P_e becomes known at each depth through the relation (11.18). The ionization conditions can be calculated, and from them we can determine the average mass per free particle m. The density follows from the perfect gas equation of state:

$$P = \frac{k}{m}\rho T \qquad (11.20)$$

We see that the assumption that the temperature-optical depth relation is known is sufficient information to calculate the entire model atmosphere.

How do we find the temperature-optical depth relation to begin with? Much effort has been applied to the two problems of finding a good first approximation and of improving a given approximation that is not good enough. These are the subjects of the next two sections.

A given T-τ_o relation is the basis of a trial model atmosphere. The test of this model is whether it is consistent with radiative equilibrium, i.e., whether the radiative flux has the correct value at all depths. The amount of calculation in the test is quite large, but it is straightforward and easily carried out with computors.

It will be noted that three different quantities had to be known or assumed before the model atmosphere could be calculated: T_e (or F); g; and the abundances z_i. These are the parameters of a model stellar atmosphere, and they uniquely determine the structure of the atmosphere, including the radiation field, within the framework of the physical theory used. If this physical theory is accurate, then the model atmospheres should show in detail the circumstances of real stellar atmospheres. If not, the numerical details will not represent real stars; however, the accuracy of the physical theory, including the assumption of LTE, radiative equilibrium, plane geometry, accurately known absorption coefficients, etc., does not affect the general conclusion that the entire structure of a stellar atmosphere is determined uniquely by the values of the three parameters T_e, g, and z_i. Exceptions to this statement are provided by such things as a large magnetic field, rapid rotation, and outside influences, as for close binary stars.

12. Reduction to the Gray Solution

One of the methods for trying to find a good first approximation for the $T(\tau_o)$ relation consists of trying to reduce the non-gray problem so that it has the gray problem solution. The fact that real stars are not too far from being gray bodies gives a basis for expecting some degree of success.

The opacity κ_o still has not been specified. The question is whether the opacity can be defined in such a way that $T(\tau_o)$ is the same relation as the gray temperature distribution. Since the latter is known, such a way of defining the opacity would make the temperature-optical depth relation known, and our problem would be solved, as indicated in the last section.

The equation of transfer at frequency ν is

$$\frac{\mu}{\kappa\rho} \frac{dI_\nu}{dz} = I_\nu - B_\nu \qquad (12.1)$$

while it is desired that the opacity be such that it satisfies the gray equation:

$$\frac{\mu}{\kappa_0 \rho} \frac{dI}{dz} = I - B \qquad (12.2)$$

One can arrive at equation (12.2) from (12.1) by making the opacity satisfy

$$\frac{1}{\kappa_0} \frac{dI}{dz} = \int_0^\infty \frac{1}{\kappa} \frac{dI_\nu}{dz} \, d\nu \qquad (12.3)$$

This equation cannot be satisfied rigorously, as the intensity depends upon direction, while the opacity cannot if the analysis of the last section is to be valid. Also, the intensity is one of the unknowns of the problem, so equation (12.3) is not of much help in this form.

Since the intensity becomes nearly isotropic in the deeper layers, an opacity which is based approximately on the above equation is a possibility. If equation (12.1) is multiplied by κ/κ_0 and integrated over frequency, we find

$$\frac{\mu}{\kappa_0 \rho} \frac{dI}{dz} = \int_0^\infty \frac{\kappa}{\kappa_0} (I_\nu - B_\nu) \, d\nu$$

$$= I - B + \int_0^\infty (\frac{\kappa}{\kappa_0} - 1)(I_\nu - B_\nu) \, d\nu \qquad (12.4)$$

If we could make the last term above vanish, then our problem would be solved. Since this cannot be done in general, we will see if we can make this term very small:

$$\int_0^\infty (\frac{\kappa}{\kappa_o} - 1)(I_\nu - B_\nu)\, d\nu \ll I - B \qquad (12.5)$$

One might think of trying to satisfy this inequality on the average, by replacing the intensity by the mean intensity; however, this could not be satisfied in radiative equilibrium as can be seen from equation (6.12).

The equation of transfer can be written in the following way:

$$\begin{aligned}
I_\nu &= B_\nu + \mu\frac{dI_\nu}{d\tau} \\
&= B_\nu + \mu\frac{d}{d\tau}(B_\nu + \mu\frac{dI_\nu}{d\tau}) \\
&= B_\nu + \mu\frac{dB_\nu}{d\tau} + \mu^2\frac{d^2 B_\nu}{d\tau^2} + \ldots \qquad (12.6)
\end{aligned}$$

The first term on the right side of equation (12.6) is isotropic, while the others depend upon direction. Since the intensity tends toward isotropy at large depths, it follows that the first term must dominate at great depths. The intensity approaches the Planck function.

If only the first two terms of equation (12.6) are used, the opacity can be defined so that the left side of equation (12.5) vanishes. One finds

$$\frac{1}{\kappa_o}\frac{dB}{dz} = \int_0^\infty \frac{1}{\kappa}\frac{dB_\nu}{dz}\, d\nu \qquad (12.7)$$

This is simply equation (12.3) with the intensity replaced by the Planck function. The derivative dB_ν/dz is not known, as it depends on the structure of the atmosphere; however, we can multiply this equation by dz/dT and cancel the common factors. This changes the position derivative to a temperature derivative of the Planck function, which does not depend on the particular model being considered. The result is the following expression for the opacity:

$$\frac{1}{\kappa_0} \frac{dB}{dT} = \int_0^\infty \frac{1}{\kappa} \frac{dB_\nu}{dT} \, d\nu \qquad (12.8)$$

The opacity defined by this relation is known as the Rosseland mean absorption coefficient.

For points which are not too close to the surface, the use of the Rosseland mean leads to a $T(\tau_0)$ relation which is the same numerical function as the gray temperature distribution, a known function. This is why the Rosseland mean is universally used in studies of the interiors of stars. The atmosphere is near the surface and the first two terms of the expansion (12.6) are not a very accurate representation of the intensity; nevertheless, the use of the Rosseland mean allows one to obtain a reasonably good first approximation to the correct temperature distribution. This first approximation is sufficient for many purposes, but if higher accuracy is wanted a method of correcting a given distribution is used. Methods of making this correction are discussed in the next section.

If the expansion (12.6) is used to evaluate the flux, it is seen that only terms with odd powers of μ make a contribution. The 2 (or 3) term expansion yields

$$F_\nu = \frac{4\pi}{3} \frac{dB_\nu}{d\tau} = \frac{4\pi}{3\kappa\rho} \frac{dB_\nu}{dz} = \frac{4\pi}{3\kappa\rho} \frac{dT}{dz} \frac{dB_\nu}{dT} \qquad (12.9)$$

The flux depends directly on deviations of the intensity from isotropy. If dB_ν/dT from the above is substituted into equation (12.8), we find the following alternative expression for the Rosseland mean:

$$\kappa_0 = \frac{1}{F} \int_0^\infty \kappa F_\nu \, d\nu \qquad (12.10)$$

This is identical to equation (11.8); thus, the Rosseland mean also allows the radiation pressure to be expressed in the simplified form of equation (11.6), in a certain approximation.

Other forms of the opacity have been introduced in order to improve upon the accuracy of the Rosseland mean for stellar atmospheres. The most important of these is the Chandrasekhar mean, given by an expression similar to equation like (12.10) except that gray fluxes are used instead of the actual fluxes. The Chandrasekhar mean is designed specifically for the atmosphere; it does seem to improve slightly on the accuracy of the Rosseland mean, but it also increases the difficulty of the calculations. In view of the efficient methods now available to improve on temperature distributions, it is perhaps best to simply use the absorption coefficient at some reasonable frequency for the opacity.

13. Corrections to the Temperature Distribution

If the use of the appropriate opacity does not lead to a model atmosphere with the correct fluxes to the desired accuracy, then a correction procedure must be resorted to. The fluxes of the trial model have been calculated, so the sizes of the errors are known. The problem is to determine the necessary changes in temperature at each depth to correct the fluxes by the desired amounts.

The problem is complicated by the fact that the flux at a given depth depends on conditions both above and below the given layer. Suppose that the temperature of a thin layer is increased by a small amount, other layers remaining the same. The energy output of that layer is increased, so that higher layers will have a larger upward flux, while the flux will be decreased in lower layers. The amounts of these changes will depend upon the optical distance from the thin zone with the temperature change as well as on the thickness of that zone. All of these optical distances depend upon frequency through the variations of the absorption coefficient. The problem is further complicated by the fact that most of the obvious simplifications one might think of are valid only at large depths, and so are of limited help in the higher layers of an atmosphere.

For references to a number of different correction procedures, one can consult J. C. Pecker, Ann. Rev. Astron. & Astrophys. 3, 135, 1963, or the Mihalas book. Here I wish to only illustrate with a method which is quite simple but which, in practice, does not work very efficiently.

For a gray atmosphere, equations (6.12) and (6.13) can be integrated over frequency; the results are

$$\frac{dF}{d\tau} = 4\pi(J - B) \tag{13.1}$$

$$\frac{dP_r}{d\tau} = \frac{F}{c} \tag{13.2}$$

If we integrate equation (13.2) over optical depth, we find

$$P_r(\tau) = \frac{1}{c} \int_0^\tau F(\tau')d\tau' + C \tag{13.3}$$

If we now use the third relation of (8.3) and the approximate boundary condition (8.6), equation (13.3) changes to

$$J(\tau) \simeq \frac{3}{4\pi} \int_0^\tau F(\tau')d\tau' + \frac{F_o}{2\pi} \tag{13.4}$$

Now solve equation (13.1) for $B(\tau)$ and substitute (13.4) for J:

$$B(\tau) \simeq \frac{F_o}{2\pi} + \frac{3}{4\pi} \int_0^\tau F(\tau')d\tau' - \frac{1}{4\pi} \frac{dF(\tau)}{d\tau} \tag{13.5}$$

If the temperature at each depth is changed by the amount δT, then this will produce a change in the Planck function of δB and a change in the flux of δF; from equation (13.5) we see that these changes are approximately related by

$$\delta B(\tau) \simeq \frac{\delta F_o}{2\pi} + \frac{3}{4\pi} \int_0^\tau \delta F(\tau')d\tau' - \frac{1}{4\pi} \frac{d}{d\tau}\delta F(\tau) \tag{13.6}$$

The flux corrections $\delta F(\tau)$ are known from the trial model, and so $\delta B(\tau)$ and, finally, the temperature corrections $\delta T(\tau)$ can be found. Although grayness was assumed for the atmosphere, the corrections will also apply roughly to a non-gray model if the departures from grayness are not too large. The method described above does not work well in practice and was given for illustrative purposes only; methods which do work well are considerably more complicated. Using one of the better methods one can construct a model atmosphere which satisfies the flux conditions to an arbitrarily high accuracy.

The final result of all of these calculations is a detailed knowledge of the physical conditions in the model atmosphere; in many cases we are probably justified in believing that this means knowledge of conditions in the atmospheres of real stars. Table 5 is an example of a model of the solar atmosphere calculated by R. L. Kurucz, Ap. J. Suppl. 40, 1, 1979 This paper contains a very large number of model atmospheres of stars covering a range of effective temperatures, surface gravities, and abundances. See problem 13 for a simpler approach.

Table 5

Model Solar Atmosphere

τ_o	z (km)	T	P	N_e	ρ	κ_o	F_c/F
0.00	0	-	-	-	-	-	0
0.01	270	4709	1.29+4	1.58+12	4.36-8	0.041	0
0.03	337	4850	2.37+4	2.85+12	7.79-8	0.070	0
0.10	405	5069	4.36+4	5.40+12	1.37-7	0.12	0
0.32	477	5467	7.92+4	1.18+13	2.31-7	0.22	0
1.0	541	6301	1.29+5	5.47+13	3.26-7	0.75	0.01
3.3	576	7404	1.62+5	3.60+14	3.48-7	3.7	0.48
10	605	8209	1.90+5	1.12+15	3.68-7	10.5	0.83

The units in Table 5 are all cgs. N_e is the electron density, and κ_o is the Rosseland mean. The last column gives the fraction of the total flux which is carried by convection. This is the subject of the next section.

14. Convection

The combination of plane geometry and energy conservation requires that the total flux be constant through the atmosphere. We have previously assumed that radiation is the only important means of carrying energy out of a star, so the radiative flux was constant with depth. We now allow for the possibility of convective motions in the gas which can carry energy; thus radiative equilibrium may no longer hold.

If F_r and F_c are the radiative and convective fluxes, then the new condition in a plane atmosphere is

$$F_r + F_c = \sigma T_e^4 \qquad (14.1)$$

F_r is calculated as before, but now F_c needs to be considered.

The presence of convection means, of course, that there are large scale mass motions in the gas. Hydrostatic equilibrium cannot be exactly realized, and the equations of hydrodynamics may have to be introduced. The motions are usually slow enough, however, that hydrostatics remains a very good approximation. In the solar atmosphere, for example, the gas pressure is about 10^5 cgs where the density is about 10^{-7} cgs. The convective velocities are observed to be around one km/sec and so the hydrodynamic pressure $\rho v^2/2$ is no more than about a percent of the gas pressure. For purposes of further discussion and without much loss of physical reality, we will assume that hydrostatic equilibrium is satisfied.

First we will establish the conditions under which convection will occur. Consider a mass element in equilibrium with its surroundings. Suppose that a random perturbation causes this element to be displaced downward by a small amount. The element is in new surroundings with different physical conditions, and it will feel a force on it. If the force is exerted upward, tending to push the element back where it came from, the material is stable against convection: motions are damped out and convection does not occur. If the force is downward, tending to push the element further away from its original position, the material is unstable against convection: random motions are enhanced and convection does occur.

Let P, T, and ρ be the original pressure, temperature, and density of the element. If δz is the distance the element is displaced downward, then the new density of the surroundings is $\rho + \delta z (d\rho/dz)$, where a derivative given without a subscript means that it is the actual gradient in the atmosphere. If we assume that the element quickly contracts to have the same pressure as its surroundings, and if energy exchanges are not important during this contraction, then the new density will be $\rho + \delta P (d\rho/dP)_{ad}$, where the subscript ad means that the process takes place adiabatically, and where δP is the pressure change in the stellar material along δz. The element sinks further if it is more dense than its new surroundings, so the condition for convection is

$$\left.\frac{d\rho}{dP}\right)_{ad} > \frac{d\rho}{dP} \qquad (14.2)$$

If the density in the atmosphere increases less rapidly than the adiabatic gradient, the lower layers are not able to support the upper ones without becoming unstable to convective motions. The equation of state allows this condition to be expressed in terms of the temperature; for a perfect gas, we have

$$\frac{d}{dP}\left(\frac{T}{m}\right) > \left.\frac{d}{dP}\left(\frac{T}{m}\right)\right)_{ad} \qquad (14.3)$$

where m is the average mass per free particle. If m is nearly a constant, this reduces to the more familiar form

$$\frac{dT}{dP} > \left.\frac{dT}{dP}\right)_{ad} \qquad (14.4)$$

as the condition for convection to exist.

If the temperature gradient is superadiabatic, convection occurs. In constructing a model atmosphere, one must calculate the adiabatic gradient in order to make the convection test. For a perfect gas which is completely neutral or completely ionized, $dT/dP)_{ad} = 0.4T/P$. When the material is partly ionized, this gradient is a complicated function of the physical conditions. Unsöld derives this expression in his book Physik der Sternatmosphären.

When one constructs a model atmosphere without the restriction of radiative equilibrium, one must check at each point to see if the inequality (14.4) is satisfied. If not, the convective flux is zero at that point. (Actually, there can be some overshooting of convection into the neighboring region.) If the inequality is satisfied, then one must have a method of calculating F_c so that the condition (14.1) can be tested.

The convective flux is usually calculated by the highly idealized mixing length theory. It is supposed that the convective medium consists of a large number of small mass elements, each of which moves as a unit. The elements are continually being formed, they move up or down a certain distance and then dissipate back into the background. The average vertical distance they move before dissipation is known as the mixing length L. There is an average temperature at each depth which is somewhat cooler than the rising elements and somewhat hotter than the falling elements.

Let T, P, and ρ be the average values of the variables at a certain depth. Let $T + \delta T$, $\rho + \delta \rho$ be the temperature and density in typical elements at the same depth. For rising elements, δT will be positive and $\delta \rho$ will be negative; the opposite will hold for sinking elements. It is generally assumed that the elements have expanded or contracted until they are at the same pressure as the surroundings. If C_p is the specific heat per unit mass at constant pressure, then $C_p \rho \delta T$ is the excess energy per unit volume (positive for rising elements, negative for falling ones) in the elements. Thus the vertical motions cause a net upward transfer of energy. Let v be the mean upward velocity of the elements; then

$$F_c = C_p \rho v \delta T \qquad (14.5)$$

The velocity and temperature difference must now be found in terms of the local variables.

The buoyant force exerted on a mass element per unit volume is $g\delta\rho$, where g is the acceleration of gravity. The work done on the element is then

$$W = \int_0^{L/2} \delta\rho g \, dz \qquad (14.6)$$

An element chosen at random has been accelerated over the distance of L/2. Now if the density difference is taken as proportional to the displacement for convenience (the final result is not sensitive to this assumption), the above becomes

$$W = \frac{1}{4} gL\delta\rho \tag{14.7}$$

Since there is no pressure difference between element and the surroundings, the perfect gas law gives $\delta\rho = -\rho\delta T/T$. We can express δT in terms of the difference in temperature gradients between element and the general medium; if there is little energy exchange during the life of an element, the gradient within an element will be the adiabat. Finally, we can eliminate the velocity from equation (14.5) by equating equation (14.7) with the mean kinetic energy per unit volume in an element. All of these substitutions lead to the following equation for the convective energy flux:

$$F_c = c_p\rho \left(\frac{g}{T}\right)^{1/2} \frac{L^2}{4} \left[\frac{dT}{dz} - \left(\frac{dT}{dz}\right)_{ad}\right]^{3/2} \tag{14.8}$$

The mixing length theory does not determine the value of L, so it is a free parameter. One might expect it to be of the order of the scale height in the atmosphere, as the elements should lose their identity if they move into regions of very different conditions from where they originated. Another possibility is that L is about the distance to the nearest boundary, if that is less than the scale height. As the theory leading to equation (14.8) is somewhat artificial, one should not be surprised if the numerical coefficient needs to be adjusted by several factors of two. Different possible values of L are due to uncertainties in the convective theory. L is not an independent parameter of the atmosphere in the way of T_e, g, and the chemical composition.

For a more detailed treatment, one can consult E. Vitense, Zs. f. Ap. 32, 135, 1953, and E. Böhm-Vitense, Zs. f. Ap. 46, 108, 1958. The framework of the mixing length theory, however, probably places a limit on the accuracy with which the actual convection in stars can be represented by relations such as equation (14.8).

Stars which are not too hot have a convection zone in the lower part of their atmospheres. The properties of the zone are strongly influenced by the excitation and ionization of hydrogen, so it is called the hydrogen convection zone. The upper boundary is near optical depth unity, its precise position depending on the type of star. The cause of this outer convection zone can be understood from the following simplified analysis.

Suppose we have a gray atmosphere. If convection is not important, then the temperature distribution will be approximately that given by the Eddington approximation of equation (8.11). If we use this and the equation of hydrostatic equilibrium (11.14), then we find

$$\frac{P}{T} \frac{dT}{dP} = \frac{\kappa P}{4g(\tau + 2/3)} \qquad (14.9)$$

Convection will occur if this is greater than the corresponding adiabatic value, which is 0.4 if ionization is not changing much with depth. At the surface, P and τ go to zero and the layers are not convective. What about the deeper layers?

Suppose that the absorption coefficient κ varies roughly as P^n. For large positive n, the absorption increases strongly with depth. For this assumed absorption, the hydrostatic equilibrium relation yields $\kappa P = g\tau(n + 1)$. It follows that convection occurs if

$$\frac{\tau}{\tau + 2/3} > \frac{1.6}{n + 1} \qquad (14.10)$$

This relation cannot be satisfied at any depth unless n is greater than about 0.6. Thus the absorption coefficient must increase with depth faster than this in order for an outer convection zone to occur. The more rapid the increase of the absorption with depth, i.e., the greater n, the closer to the surface the convection zone comes. While this example is very artificial, it does illustrate some important properties of convection zones.

If the convective flux is significant, then radiative equilibrium is no longer valid. Convection will tend to lower the temperature gradient, as radiation will then need to carry less energy

In the atmospheres of moderate temperature stars, the dominant form of absorption is the negative ion of hydrogen. The formation of this ion requires the collision of a free electron and a neutral hydrogen atom, so the abundance of H^- is very sensitive to the density. This causes κ to increase rapidly with depth and, as we have seen, it favors convection.

A second cause is that at lower temperatures the n = 2 level of neutral hydrogen is very sensitive to the temperature in its population, and absorption from this level is very strong in the near ultraviolet. As a secondary effect the ionization of hydrogen lowers the adiabatic temperature gradient somewhat and, as we see from equation (14.4), this also favors convection.

The occurrence of convection does not necessarily rule out radiative equilibrium. The very low material densities in stellar atmospheres make convection an inefficient carrier of energy. The temperature gradient can be considerably greater than the adiabat without F_c being very appreciable. In the deeper layers convection becomes very efficient, and only a very small difference between the actual and the adiabatic gradients is sufficient to carry essentially all of the flux. This region, however, usually occurs well below the atmosphere.

We now examine the effects of a convection zone in the simplified case of the Eddington approximation to the gray atmosphere. Let convection occur below the optical depth τ_o, and the radiative flux is then given by

$$F_r = F_o \qquad \tau \leq \tau_o$$
$$F_r = F_o - F_c \qquad \tau > \tau_o \qquad (14.11)$$

The total flux is denoted by F_o to emphasize that it is a constant with depth. Then the Eddington approximation of equation (8.4) along with the boundary conditions (8.6) and (8.7) lead to the following expression for the mean intensity:

$$J(\tau) = \frac{F_o}{2\pi} + \frac{3}{4\pi} \int_0^\tau F_r(\tau) \, d\tau \qquad (14.12)$$

We can now relate the source function to the mean intensity through equation (6.12); remember that F in earlier expressions is the radiative flux, not the total flux. We have

$$B(\tau) = J(\tau) - \frac{1}{4\pi} \frac{dF_r}{d\tau} = J(\tau) + \frac{1}{4\pi} \frac{dF_c}{d\tau} \qquad (14.13)$$

We now substitute equation (14.12) and carry out the integration as far as possible; then replacing the Planck function with temperature and the flux with the effective temperature, we find

$$T^4 = \frac{3}{4} T_e^4 (\tau + \frac{2}{3}) \qquad \tau \leq \tau_0$$

$$T^4 = \frac{3}{4} T_e^4 (\tau + \frac{2}{3} - \int_{\tau_0}^{\tau} f_c d\tau + \frac{1}{3} \frac{df_c}{d\tau}) \qquad \tau > \tau_0$$

(14.14)

Here $f_c = F_c/F_0$ is the fraction of the total flux carried by convection.

We see from the first relation of equation (14.14) that the upper, radiative layers are not affected by the convection zone in the Eddington approximation; this is the same as equation (8.11) for radiative equilibrium. This is obviously not correct physically but a result of the approximation used, as one cannot adjust the temperature in one region without affecting all regions. We do expect the upper layers to be affected, but by a smaller amount than the layers that are in the convection zone.

We see from the second equation of (14.14) that there are two terms which are introduced by the effects of convection. The first one is always negative, so it has the effect of lowering the temperature as compared with the case of radiative equilibrium. If the radiative flux were constant at its surface value, then the temperature would have to increase as indicated by the first two terms. The presence of a convective flux means a lowering of the radiative one, and this is reflected in the integral term.

The final term in equation (14.14) involves only the derivative of the convective flux. If the radiative flux is decreasing with depth, which means the convective flux is increasing, then there must be an excess of energy emitted over energy absorbed, which translates into an excess of the source function over the mean intensity. See equation (6.12). Note that this derivative term needs to vanish at the convective boundary in order for the temperature to be continuous.

Figure 5 shows the effects of convection on the temperature distribution. The ordinate is $t = 4T^4/3T_e^4$, while the

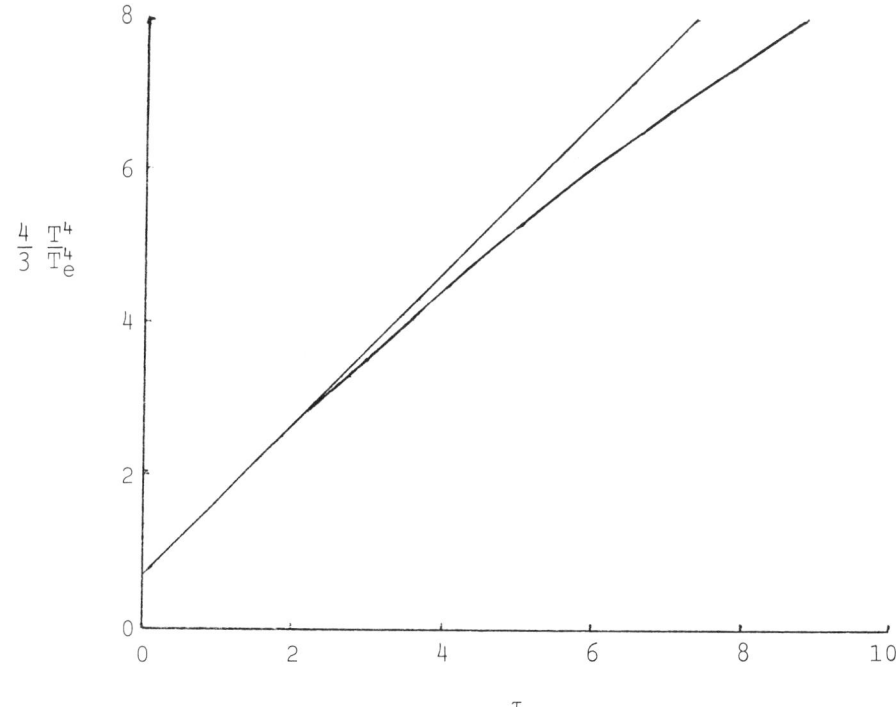

Figure 5. The effects of convection on the temperature distribution in the Eddington approximation. The upper curve is for radiative equilibrium, and the lower curve has a convection zone beginning at $\tau_o = 1$. See the text for a description of the convective flux.

abscissa is optical depth. The convective flux was artificially set by the simple assumption $f_c = 0.05(\tau - 1)$, with the top of the convection zone being at $\tau = 1$. Note that this expression does not satisfy the continuity conditions at the upper boundary, but the deviations are small and not important in the context of the Eddington approximation. The cooling effects of the convection are apparent. It should be noted, however, that the form assumed here is considerably more efficient than what one would generally expect in practice.

Numerical calculations and observations show that convection has only a small effect on the atmospheric structure and emitted radiation of most stars. Unless there is particular interest in the relatively deep layers of an atmosphere, one can retain the assumption of radiative equilibrium for most types of stars.

15. Semi-Empirical Models

The methods described previously for constructing model atmospheres are entirely theoretical, but a method exists which uses observations directly. Unfortunately, the detailed observations necessary are available only for the Sun, which does place a decided limitation on the method; however, it does provide an important an important check on the purely theoretical models.

For a semi-infinite, plane atmosphere in LTE, the intensity at the surface is

$$I_\nu(0,\mu) = \int_0^\infty B_\nu(\tau) \, e^{-\tau/\mu} \, \frac{d\tau}{\mu} \qquad (15.1)$$

This relation indicates that the depth dependence of the Planck function, which means the depth dependence of the temperature, is related directly to fixing the variation of intensity with direction. If $T(\tau)$ is known, then $I_\nu(0,\mu)$ is easily obtained. It is pointed out in Section 11 that $T(\tau)$ is all that is needed to construct a model atmosphere. Since the surface intensity is observable, the problem then presents itself of finding if equation (15.1) can be inverted. Can observations of the intensity be used to find the temperature-optical depth relation needed to construct a model atmosphere?

Direct attempts at a mathematical inversion of equation (15.1) have not been particularly successful. An alternative is to express the Planck function as an expansion in optical depth with adjustable parameters; these parameters are then forced to have values that give a satisfactory fit to the observed intensity. A number of investigators have constructed semi-empirical models of the Sun in this way, the most notable being by A. K. Pierce and J. H. Waddell, Mem. R. A. S. 68, 89,

1961.

Let B_ν^* be the ratio $B_\nu/I_\nu(0,1)$, where $I_\nu(0,1)$ is the intensity at the center of the disk. Then assume that B_ν^* has an expansion as follows:

$$B_\nu^*(\tau) = a_1 f_1(\tau) + a_2 f_2(\tau) + a_3 f_3(\tau) + \ldots \quad (15.2)$$

The a_i are constants at a given frequency. If this is put into equation (15.1), we find

$$I_\nu^*(0,\mu) = a_1 F_1(\mu) + a_2 F_2(\mu) + a_3 F_3(\mu) + \ldots \quad (15.3)$$

Here the $F_i(\mu)$ are the transform functions of the $f_i(\tau)$ through equation (15.1).

Limb darkening observations of the Sun show that $I_\nu^*(0,\mu)$ is with fairly high accuracy a linear function of μ. This means that $F_1 = 1$, $F_2 = \mu = \cos\theta$. We find from inspection that $f_1 = 1$, $f_2 = \tau$. A number of possible functions for f_3 have been tried, the most popular being the second exponential integral function (see equation (6.6)). With this we find the following expressions for the 3 term expansions:

$$B_\nu^*(\tau) = a_1 + a_2\tau + a_3 E_2(\tau) \quad (15.4)$$

$$I_\nu^*(0,\mu) = a_1 + a_2\mu + a_3\left[1 - \mu\ln\left(1 + \frac{1}{\mu}\right)\right] \quad (15.5)$$

Pierce and Waddell indicate that the three term expansion fits the observed limb darkening to within observational errors, so additional terms will not improve the accuracy.

A set of expansion parameters a_i is found by a least squares fit to the observations at a given frequency. This determines the values of B_ν^*; if the intensity values at the center of the disk are known, the Planck function and the temperature become known functions of optical depth at the same frequency. Enough information is known to construct a model atmosphere without resorting to the methods of Sections 12-14.

There is another important property of the limb darkening observations. When we obtain the temperature-optical depth relation at a given frequency, we can differentiate it to find $d\tau/dT = \kappa\rho dz/dT$, where z is geometrical distance into the star.

Let us find this derivative at two separate frequencies, but at a fixed temperature, which means a fixed value of z and of ρ. Then by taking the ratio of these, we have evaluated

$$\frac{d\tau(\nu_1)}{d\tau(\nu_2)} = \frac{\kappa(\nu_1)}{\kappa(\nu_2)} \qquad (15.6)$$

In other words, the frequency dependence of the absorption coefficient, at any depth, is obtained directly from the observations. This provides an important check on the theoretically determined absorption coefficient of the Sun. Note that the semi-empirical method does not depend upon the assumption of radiative equilibrium.

The semi-empirical method produces a model atmosphere which, by definition, satisfies the limb darkening observations; however, this does not mean that the model is accurate in detail. Limb darkening is known to be rather insensitive to many aspects of a model atmosphere. The three term expansion of equation (15.4) may be accurate for optical depths in the range of 0.5-1.5 or so, but it can be very inaccurate if applied to layers far from this range. The semi-empirical model provides an important check on the solar atmosphere, but it does need supplementing by other methods.

16. Continuous Absorption and Blanketing

In order to construct a model atmosphere, it is necessary to be able to calculate the absorption coefficient as a function of the chemical abundances, frequency, and the physical conditions. The important absorbing agents must be identified, and the absorption cross sections must be available from calculations or from laboratory data. It is not the purpose here to reproduce either the quantum mechanical calculations or a long list of tabulations. The general references may be consulted for this. Here only a few general comments will be made on the subject.

It has already been mentioned that the main source of continuous absorption in the atmospheres of moderate temperature stars is the negative ion of hydrogen H^-. This ion has only one bound stars, which is about 0.75 eV below the continuum. Photons of wavelength less than about 16,500 Å can ionize H^- to produce neutral hydrogen plus a free electron. H^- never becomes very abundant: at higher temperatures it is too easily ionized, while at lower temperatures there is only a very small source of free electrons to provide for it. In spite of the low abundance, it dominates the absorption coefficient. For wavelengths longer than 16,500 A, only free-free transitions are possible. These consist of the acceleration of a free electron which is temporarily in the field of a nearby neutral hydrogen atom.

One of the interesting consequences of H^- dominating the absorption coefficient is that the coefficient does not vary strongly with frequency. The bound-free coefficient varies by less than a factor of two between 4000 A and 13,000 A; as a result, gray model atmospheres are good approximations to the real thing.

At other temperatures other absorbers are important. For moderately hot stars neutral hydrogen is dominant, while neutral and ionized helium as well as electron scattering become important at very high temperatures. At lower temperatures molecules have an increasingly important effect. There are many sources of absorption which, while not very important, neither are they weak enough to neglect. There are also some absorbers which are important over limited wavelength ranges, while having negligible effect on the overall structure of the atmosphere. The details can be many and complicated.

Bound-bound transitions also have an effect on the emitted radiation and on the structure of an atmosphere. A single line has such a narrow width that it generally has no significant effect on the energy balance of the atmosphere; however, the cumulative effect of large numbers of such lines is of importance, and a statistical way of taking them into account needs to be used. This combined effect of all of the lines is known as the blanketing effect, a name derived from the fact that the lines act like a blanket, making it more difficult for heat to escape.

In the simplest blanketing model, the so-called picket fence representation, the absorption lines are assumed to have square profiles. The main parameter is one which fixes the relative probabilities that, at any given wavelength, one is in a line or between lines. In more sophisticated treatments, one calculates the line absorption coefficient at a given wavelength from data on the positions and shapes of a very large number of nearby lines. It would appear that, as far as the structure of the atmosphere is concerned, the simple theories are about as accurate as the more complicated ones; however, for an accurate prediction of the frequency distribution of the intensity or the flux at high resolution, the more complicated approaches are necessary, of course.

Blanketing is not very well represented by simply adding a slowly varying function of wavelength to the continuous absorption coefficient. Blanketing causes a small wavelength interval to receive radiation from both high and low regions in the atmosphere, and replacing this range of contributions by a single intermediate depth of formation does not give the same effect.

The absorption lines in stellar spectra do not occur uniformly at all wavelengths, but are much more prevalent at the shorter wavelengths. This causes a distortion of the emitted radiation which affects the color of the star. This is a very pronounced effect, and it complicates efforts to determine information from the observed colors. If the temperatures of stars are to be compared through their colors, one must make sure that they have equal amounts of blanketing. Most of the blanketing is due to lines of the heavy elements, so this means the stars should have the same chemical abundances. Since there are in fact significant compositional differences among stars, it is necessary to correct for differential effects of blanketing. See Sandage and Eggen, M.N.R.A.S. 119, 278, 1959; Wildey, Burbidge, Sandage, and Burbidge, Ap.J. 135, 94, 1962, where these questions are considered.

Chapter 4: Line Formation

17. Line Absorption and Emission

A spectral line results when the absorbing and emitting properties of a medium change significantly over a very small frequency interval. The frequency range of a line is usually small enough that the properties of the continuum can be considered as constant over the line. The line represents a transition between two bound levels of an atom.

Much of the material in the earlier chapters applies equally well to line radiation and to continuum radiation. One can solve for the line radiation when the line absorption and emission coefficients, or equivalently, the absorption coefficient and the source function, are known throughout the medium. One simplification that usually applies to the lines but not to the continuum is that the energy in one line is too small to appreciably affect energy balance in the medium. Thus one can study the formation of a given line without the necessity of a simultaneous solution for a model atmosphere. It is usually assumed that a model atmosphere is already avaliable when a line study is made. This need not be the case, however, if we are studying very strong lines which are formed far above the regions important for the continuum.

It is convenient to introduce Einstein coefficients for the line processes as was done for the continuum in equations (5.12). In fact the only modification needed is that the right sides should all be multiplied by the profile function ϕ_ν. This is defined so that $\phi_\nu d\nu$ is the probability that, if the given transition does take place and if the intensity is constant across the line, the absorbed or emitted photon has a frequency between ν and $\nu + d\nu$. There is actually a separate profile function for absorption and for emission, but the differences between them are generally small and will not be considered here. See the books by Jefferies and by Mihalas for further details on this.

The profile function satisfies the condition

$$\int_0^\infty \phi_\nu d\nu = 1 \tag{17.1}$$

It is written as a separate factor in the definitions of the Einstein coefficients so that these coefficients can be considered as atomic constants, independent of physical conditions.

With the above modification, the analysis following equations (5.12) is valid for line radiation. In particular, the absorption coefficient is given by equation (5.15) multiplied by the profile function. If a is the atomic absorption coefficient, given by k/N_1, then from equation (5.15) we have

$$a = \left(1 - \frac{N_2 B_{21}}{N_1 B_{12}}\right) \frac{h\nu}{4\pi} B_{12} \phi_\nu \tag{17.2}$$

The units of this are cm²/atom in the lower level.

It is common to express the transition probability in terms of the oscillator strength or f-value. For the transition from level 1 to level 2, this is given by

$$f_{12} = \frac{mch\nu}{4\pi^2 e^2} B_{12} = \frac{g_2}{g_1} \frac{mc^3}{8\pi^2 e^2 \nu^2} A_{21} \tag{17.3}$$

Equations (5.19) still hold between the different Einstein coefficients. In terms of the f-value, the atomic line absorption coefficient in equation (17.2) is

$$a = \left(1 - \frac{g_1 N_2}{g_2 N_1}\right) \frac{\pi e^2}{mc} f_{12} \phi_\nu \qquad (17.4)$$

Note that ϕ_ν is the only part of the absorption coefficient that varies with frequency across the line. The first factor on the right side of equation (17.4) is the correction for induced emission. The absorption coefficient to be used in the equation of transfer is the net coefficient: the excess of absorptions over induced emissions. If the latter are more abundant than absorptions, the effective absorption coefficient is negative and there is an exponential increase of radiation, as in a laser. Near thermal equilibrium (TE), the population ratio of the levels is given by equation (5.18), and the above term becomes $(1 - e^{-h\nu/kT})$. The value of the oscillator strength must be determined from calculations or from laboratory experiments. The profile function is the subject of Section 19.

Since the source function is the ratio of emission and absorption coefficients, the profile function cancels out, and equations (5.17), (5.20), and (5.22) are valid for line transitions. As before, if conditions are sufficiently close to TE that the excitation temperature for the two levels, defined by equation (5.21), is very close to the kinetic temperature throughout the line forming region, then the given line is formed under LTE (local thermal equilibrium) and the source function is the Planck function of the kinetic temperature. Lines are usually formed higher in the atmosphere than the continuum, so we can expect on general grounds that deviations from LTE are greater for lines than for the continuum.

For a given atomic level j, it is common to define a departure coefficient b_j through the relation

$$N_j = b_j N_j^* \qquad (17.5)$$

where N_j^* is what the population of level j would be in TE at the same kinetic temperature, electron density, and ion density of the given element. Thus $b_j = 1$ in TE, and this is also true more generally for continuum levels if a Maxwellian velocity distribution prevails. The departures of the b_j's from unity are measures of departures from the Boltzmann and Saha populations. Quite generally we have

$$\frac{N_1}{N_2} = \frac{b_1}{b_2} \frac{g_1}{g_2} e^{h\nu/kT} \qquad (17.6)$$

where $\nu = E_{12}/h$, and T without a subscript is the kinetic temperature. If this is substituted into equation (5.20), we find for the line source function

$$S_{12} = \frac{2h\nu^3/c^2}{(b_1/b_2)e^{h\nu/kT} - 1} \tag{17.7}$$

The frequency subscript of the source function is left off for convenience. If $b_1 = b_2$, then S_{12} reduces to the Planck function of the kinetic temperature, and LTE holds in the line. In general this will not be the case, and one must determine the ratio (b_1/b_2) throughout the line forming region. In the next section this ratio will be found in the simple case of an atom which has only two levels. The generalization to the case of a many level atom is then briefly described. The profile function is discussed in the following section, and then we will be in a position to solve the line formation problem.

18. The Two-Level Atom

It is indicated above that if the populations of the different atomic energy levels are the same as in TE, all of the departure coefficients are unity and LTE is valid. The populations of the levels are determined by the transition rates between the levels. These transitions can be radiative or collisional, meaning that the energy of the transition can be supplied by or given to either another particle or a photon.

The collisions which take place between particles in a gas are of two kinds: elastic and inelastic. Elastic collisions exchange kinetic energy between particles, but the total amount of kinetic energy is conserved. Inelastic collisions exchange kinetic energy with other forms of energy, such as excitation and ionization energies. Elastic collisions tend to make the gas behave as in TE, as they tend to set up a Maxwellian distribution of velocities. Inelastic collisions destroy this distribution since they preferentially exchange energies which are at or above certain threshholds. In strict TE each inelastic collision is exactly balanced by its inverse, so they have no net effect. In most cases of astrophysical interest the elastic collisions are very much more common than the inelastic, so a near Maxwellian distribution is set up; degeneracy is the main exception to this.

The above would not be true if the mean free path

for the particles is not small compared to the distance over which the temperature changes appreciably. In this case, particles representing different temperatures are mixed together, and the result is not the same as in TE at a fixed temperature. It is easy to show, however, that the collisional mean free paths are indeed very small in the above sense. If we assume the Eddington approximation to the gray temperature distribution, equation (8.11), then we find

$$\frac{1}{T}\frac{dT}{dz} = \frac{k}{4(\tau + 2/3)} \qquad (18.1)$$

Near the surface the absorption coefficient k is of the order of 10^{-7} cm^{-1} for the solar atmosphere, so a change in depth of some 10^7 cm is necessary for the temperature to change by an appreciable fraction of itself. The collisional mean free path is 1/Na, where N is the particle density and a is the collisional cross section. With N being around 10^{17} particles/cm^3 and a cross section of 10^{-16} cm, the mean free path turns out to be about 0.1 cm. This is 10^8 times smaller than the temperature scale, so under these conditions collisions tend to make the departure coefficients unity.

The situation is quite different for radiative transitions. Even if each point in the atmosphere emitted as if it were in perfect TE, the atmosphere is, by definition, the region where the photon mean free path is NOT small compared to the temperature scale. The photon mean free path is just the distance corresponding to optical distance of unity; according to equation (18.1), this is same as the temperature scale at the surface. In addition, all parts of the atmosphere can "see" the boundary at the surface, so the radiation field at any point is non-isotropic. Deep into the star these conditions do not hold, and LTE is an excellent approximation. In the atmosphere, the radiative transitions can cause appreciable deviations from LTE.

We will now become more quantitative. In order to keep the algebra at a reasonable level, we will consider an atom with only two bound levels. The generalization to a many level atom is straightforward. This analysis was first carried out by R. N. Thomas, Ap.J. 125, 260, 1957.

Unless the properties of the star in question are changing rapidly with time, the population of each level remains constant with time. This means that the number of transitions 1→2 equals the number 2→1. This, of course, would not necessarily be true if the atom had more than two levels. The radiative transitions can be found from equations (5.12), including the profile function as a factor, by integrating them over all frequencies and directions. The total number of spontaneous transitions per volume and time is simply $N_2 A_{21}$. In the other two cases the integration over direction changes intensity to mean intensity J_ν; the result is then $N_1 B_{12} \bar{J}$ for absorptions and $N_2 B_{21} \bar{J}$ for induced emissions, where

$$\bar{J} = \int_0^\infty J_\nu \phi_\nu d\nu \qquad (18.2)$$

Let C_{12} and C_{21} be the probabilities per unit time that an atom will have a collisional transition in the given direction. Then $N_1 C_{12}$ is the number of collisional excitations per unit volume and time. Equating the total number of excitations to the number of de-excitations per volume-time leads to

$$N_1(C_{12} + B_{12}\bar{J}) = N_2(C_{21} + A_{21} + B_{21}\bar{J}) \qquad (18.3)$$

For a multilevel atom one would have to include transitions with all of the other levels, including both free and bound.

In TE there is detailed balance, and the collisional rates must balance each other; the same is true of the radiative rates. If this were not the case, then equation (18.3) would predict that the ratio N_1/N_2 would depend upon the relative importance of radiative and collisional processes, even in TE. The appeal to TE conditions then require $N_1^* C_{12} = N_2^* C_{21}$, where the N^*'s are the TE populations. But the kinetic properties of the gas are the same as if TE held, due to the efficiency of the elastic collisions; it follows that the C's are not effected by deviations from TE. Since the N^*'s satisfy the Boltzmann distribution, we find

$$C_{12} = C_{21} \frac{g_2}{g_1} e^{-u} \qquad (18.4)$$

where $u = E_{12}/kT = h\nu/kT$. We now express the Einstein B's in terms of A_{21} through equations (5.19); then equation (17.6) is used to eliminate N_1/N_2, and equation (18.4) is used to eliminate C_{12}. We finally solve equation (18.3) for the ratio of the b's, with the following result:

$$\frac{b_1}{b_2} = \frac{C_{21} + A_{21}(1 + c^2\bar{J}/2h\nu^3)}{C_{21} + A_{21}c^2 e^u \bar{J}/2h\nu^3} \quad (18.5)$$

Suppose that collisional transitions dominate over radiative ones; then C_{21} is much larger than A_{21}, and we see that equation (18.5) reduces to $b_1 = b_2$, i.e., LTE in the line. This is just what we expect, as the kinetic properties of the gas are the same as in TE.

Now consider the other extreme in which A_{21} is very much larger than C_{21}. The above then leads to

$$\frac{b_1}{b_2} = \frac{1}{W}(1 - e^{-u} + We^{-u}) \quad (18.6)$$

where W is the ratio of the average mean intensity to the Planck function of the local kinetic temperature:

$$W = \frac{\bar{J}}{B_\nu(T)} \quad (18.7)$$

In TE $W = 1$. An absorption line would tend to have W less than one, although the precise level of the continuum and the kinetic temperature are also involved. Near the surface geometric dilution reduces the value of W, as strong radiation comes only from the lower regions. This amounts to a factor of one-half at the surface. We expect W to decrease from unity as we approach the surface from below. Equation (18.6) then predicts (b_1/b_2) to be greater than unity at or just below the surface of the atmosphere: the excited level is less populated with respect to the ground state as compared to TE. This is the same effect as in interstellar space, where the great dilution of the radiation field causes practically all of the atoms to be in the ground state. The fact that the above was calculated for a two level atom means that the more complicated case in which both levels are excited cannot be predicted from this.

It is a straightforward procedure to evaluate the source function in equation (17.7). Define the quantity q as

$$q = \frac{C_{21}(1 - e^{-u})}{A_{21} + C_{21}(1 - e^{-u})} \tag{18.8}$$

Except for the factor $(1 - e^{-u})$, q is the ratio of collisional de-excitations to total de-excitations. If we define scattering in the line as spontaneous emission which was immediately preceeded by absorption, then q is essentially the probability of non-scattering. The line source function then turns out to be

$$S_{12} = qB_\nu(T) + (1 - q)\int_0^\infty J_\nu \phi_\nu d\nu \tag{18.9}$$

This relation is analogous to equation (5.23) for the continuum source function. Both expressions have a non-scattering part which is the Planck function of the kinetic temperature. Both have a scattering part which is proportional to the mean intensity. There is a major difference, however, between the two forms of the scattering term. In arriving at equation (5.23), it was assumed that the scattered photon suffers no frequency change, so the scattering is coherent. While there are processes which cause small frequency changes, they are of no importance in continuum processes because things vary only slowly and smoothly with frequency. For line formation on the other hand, ϕ_ν varies by many orders of magnitude over very small frequency intervals, and small frequency shifts between absorbed and emitted photons cannot be ignored. In equation (18.9) the scattering term is completely non-coherent, which means that the emitted photon has lost all memory of the frequency of the absorbed photon. Notice that in equation (18.9) the source function is essentially independent of frequency over the width of the line. To apply equation (5.23) to a strong absorption line is to have a source function which changes considerably across the line. The form of equation (18.9) is much closer to reality than is equation (5.23).

In the deep layers of the star $J_\nu \rightarrow B_\nu$ and $S_{12} \rightarrow B_\nu$, regardless of the value of q. As we approach the surface, J_ν begins to deviate significantly from B_ν, and the source function deviates from the Planck function, the amount depending on the value of q. Eventually we will reach regions where the material densities are so low that collisions are not important, and q will be very small. Thus the scattering term in equation (18.9) must become dominant above a certain level, and there is also a level in the atmosphere above which J_ν is significantly different from B_ν. If the line still has appreciable absorption above these levels, then LTE does not properly describe the formation of that line.

The kinetic temperature T at any point is determined by the kinetic properties of the particles within a collisional mean free path of that point. As stated earlier, this mean free path is very small on the scale of the atmosphere, so T is determined locally. Likewise, the mean intensity is determined by the photons within one photon mean free path; thus the whole atmosphere plays an important role in fixing the mean intensity at any point in the atmosphere.

There is another important consideration here: a scattering can change the frequency of a photon by only a small amount, and a photon must undergo a non-scattering absorption before it can come into equilibrium with its thermal surroundings. If q is very small, a photon must travel many mean free paths before it thermalizes. In deciding whether radiative processes favor LTE, it is this thermalization length rather than the mean free path which must be compared with the temperature scale height. LTE is a very great simplification, and this produces a tendency to apply it beyond its region of validity.

The total source function is the combination of that of the line and that of the continuum. They combine according to equation (5.2). If η is the ratio of line to continuous absorption and if the continuum is formed in LTE, then

$$S_\nu = \frac{1 + \eta q}{1 + \eta} B_\nu + \frac{\eta(1 - q)}{1 + \eta} \int_0^\infty J_\nu \phi_\nu \, d\nu \qquad (18.10)$$

When the above analysis is generalized to include all levels of an atom, then terms for the collisional and radiative transitions between level 1 and all others must be added. Terms which are both non-scattering and non-Planckian appear, and such quantities as b_1/b_i and $\bar{J}_{1j}$ occur for all values of j. The equation describing statistical equilibrium for level 1 is not sufficient to solve the problem. The equations for all of

the levels must be included and solved simultaneously. In
other words, the formation of a given line cannot be considered by itself, but one must consider the simultaneous formation of all possible lines and continua of the given atom.
Finally, since a photon cannot be reserved by an element but
is liable to be absorbed by any of several, it may be necessary to solve simultaneously for the transitions in more than
a single element.

For treatment of the more general theory one may consult the books by Jefferies and by Mihalas listed in the general references. The effects of departures from LTE are the
subject of a review article by Mihalas and Athay in Ann. Rev.
Astron. Astroph. 11, 187, 1973.

19. Line Broadening

The broadening or profile function ϕ_ν appears in the
line absorption coefficient and, if LTE is not valid, in the
source function. In this section the principal causes of
line broadening are examined and the resulting forms of ϕ_ν
are derived in approximate form. There are three main causes
of broadening which are generally important: the Doppler
effect, natural broadening, and pressure broadening. These
are considered in turn, and then the combined effects are
derived.

The Doppler effect causes the frequency of a photon to
depend on the motion of the observer. This is valid for both
line and continua transitions, but only the lines are sensitive to the small frequency shifts which are usually involved.
If speeds are small compared with c, the speed of light, then

$$\Delta\nu = -\frac{v\nu}{c} \quad (19.1)$$

v is the relative radial component of the velocity between
source and observer. The sign convention on this is that the
radial velocity is positive if the two are receeding from each
other. The difference in frequency $\Delta\nu$ causes a group of atoms
having a distribution of velocities to produce a line which
is observed to have a distribution of frequencies.

Let p(v)dv be the probability that an atom has a radial
velocity between v and v + dv. There is a one to one correlation between velocity and frequency shift as given by
equation (19.1), so

$$\phi_\nu d\nu = -p(v)dv = p\left(-\frac{c\Delta\nu}{\nu}\right)\frac{cd\nu}{\nu} \qquad (19.2)$$

If the velocity distribution results from the thermal motions of the atoms, then p(v)dv is given by the one-component Maxwellian distribution:

$$p(v)dv = \frac{1}{\sqrt{\pi}} e^{-(v/v_o)^2} \frac{dv}{v_o} \qquad (19.3)$$

where

$$v_o = \left(\frac{2kT}{m}\right)^{1/2} \qquad (19.4)$$

v_o is the most probable speed of a particle of mass m. It is usual to measure frequency shifts in terms of the so-called Doppler width D; this is defined as the Doppler shift due to an atom having a radial velocity equal to v_o:

$$D = \frac{\nu v_o}{c} = \frac{\nu}{c}\left(\frac{2kT}{m}\right)^{1/2} \qquad (19.5)$$

The Doppler width D can be measured in either frequency or wavelength units. Usually frequency units will be used here, although which is being used should be apparent from the context.

If equations (19.3)-(19.5) are substituted into equation (19.2), we find

$$\phi_\nu d\nu = \frac{1}{\sqrt{\pi}} e^{-u^2} du \qquad (19.6)$$

where

$$u = \frac{\Delta\nu}{D} \qquad (19.7)$$

If there are large scale mass motions which cause a dispersion in the radial velocities of the atoms in addition to the thermal motions, then these must also be included in the function p(v)dv. If the scale of these mass motions is large compared to the length of optical distance unity, then the motions are called macroturbulence. If the scale is small, they are known as microturbulence. The reason for this division is that the two extremes affect radiation in completely different ways. Microturbulence affects radiation in exactly the same way as do the thermal motions, except that it may have a different velocity distribution. Macroturbulence, however, affects the intensity only through a fixed frequency shift. If different emitting regions have different macroturbulent velocities, then the way the intensities combine to form the flux will be affected.

It is usual to assume that the microturbulent velocities have a Maxwellian distribution with the most probable velocity v_t. Two Maxwellian distributions combine to form a new distribution which is also Maxwellian, and the square of the new most probable velocity is the sum of the squares of the old ones. Thus equations (19.6) and (19.7) are still valid, but the Doppler width is now given by

$$D = \frac{\nu}{c}\left(\frac{2kT}{m} + v_t^2\right)^{1/2} \qquad (19.8)$$

The microturbulent velocity v_t is generally treated as a new parameter to be determined when calculations are compared with observations. Macroturbulence does not affect the shape of ϕ_ν, but it causes the profile function to be symmetric about a frequency different from ν_0, the normal line center.

The next effect to be considered is known as natural broadening: lines have a width that is not dependent upon outside influences. It is a result of the fact that energy levels have finite lifetimes. The uncertainty relation indicates that an energy level is smeared out by an amount that is of the order of $h/2\pi T$, where h is Planck's constant and T is the time available to measure the energy, i.e., the lifetime of the given level.

Consider the excited level i of an atom. Let $N_i(t_0)$ be the number of atoms per unit volume in this level at time t_0. These atoms will eventually undergo transitions to other levels. If the atoms are left undisturbed, they can only make spontaneous transitions to lower levels. If we ignore the transitions into level i from higher levels, we have

$$\frac{dN_i}{dt} = -N_i \Sigma A_{ij} \qquad (19.9)$$

where the summation is over all levels j below i. The integral gives

$$N_i(t) = N_i(t_o) \, e^{-(\Sigma A_{ij})t} \qquad (19.10)$$

The average lifetime T_i of the level is found from the above to be

$$T_i = \frac{1}{\Sigma A_{ij}} \qquad (19.11)$$

The energy spread of the level is of the order of

$$\Delta E_i \sim \frac{h}{2\pi} \Sigma A_{ij} \qquad (19.12)$$

For the ground state there are no lower levels; T_1 is essentially infinite and the level is extremely sharp.

The determination of the natural shape of an energy level was first made by V. Weisskopf and E. Wigner, Zs. f. Phys. 63, 54, 1930. The result can be expressed in terms of the probability $p_i(E)dE$ that an atom in state i has energy between E and E+dE:

$$p_i(E)dE = \frac{\delta_i}{\pi} \frac{dE/h}{(E-E_i)^2/h^2 + \delta_i^2} \qquad (19.13)$$

where the so-called damping constant is given by

$$\delta_i = \frac{1}{4\pi T_i} \qquad (19.14)$$

Consider now a transition between an initial level i and a final level j. The probability that the involved photon has frequency between ν and $\nu + d\nu$ is equal to the probability that the energy in the final state is $h\nu$ greater or less than that from which it started:

$$\phi_\nu d\nu = h d\nu \int_{-\infty}^{\infty} p_i(E) p_j(E + h\nu) \, dE \qquad (19.15)$$

Substituting in the form of equation (19.13) and carrying out the integration, we find

$$\phi_\nu d\nu = \frac{\delta}{\pi} \frac{d\nu}{(\nu - \nu_0)^2 + \delta^2} \qquad (19.16)$$

where

$$\delta = \delta_i + \delta_j \qquad (19.17)$$

The third effect is known as pressure broadening and occurs when the energy levels are disturbed by nearby particles. To determine this effect we must know the type of interaction, and from this we must be able to determine the probability that the energy levels are distorted by a given amount. The frequency shift of an absorbed or emitted photon is $(\Delta E_j - \Delta E_i)/h$, where the ΔE's are the energy shifts of the individual levels. If r is the distance of a disturbing particle from the absorbing or emitting atom, the interaction between the two is usually written in the form

$$\Delta \nu = \frac{C}{r^n} \qquad (19.18)$$

The constant C depends upon the levels and on the type of interaction. An approximate description of the different types of interaction follows.

Let the atom of interest be represented as a dipole of moment $\vec{p}$. If an external electric field $\vec{F}$ is applied, the moment becomes

$$\vec{p} = \vec{p}_0 + \alpha \vec{F} \qquad (19.19)$$

$\vec{p}_0$ is the intrinsic dipole moment of the atom, and the second term in equation (19.19) is the moment induced by the field; α is the polarizability of the atom. The energy change is then proportional to

$$\Delta E = h\Delta\nu \sim \int \vec{p} \cdot d\vec{F} = \vec{p}_0 \cdot \vec{F} + \frac{1}{2}\alpha F^2 \qquad (19.20)$$

When the disturbing particle is charged, that is, when

it is a free electron or an ion, the field is proportional to r^{-2}, and the interaction is known as the Stark effect. If the perturbing particle is neutral, its main effect is through its dipole field which falls off as r^{-3}.

Only hydrogen-like atoms have an appreciable intrinsic dipole moment, and for them the first term in equation (19.20) dominates, unless the field is extremely large. For other types of atoms, the second term is dominant. The electric fields of charged particles are much greater than those of dipoles, so the latter can usually be neglected if there is appreciable ionization. Because of its high abundance and of its possession of an intrinsic dipole moment, hydrogen is usually the only neutral atom of importance in astrophysics in broadening the lines of other atoms.

The above information can be combined with equations (19.18) and (19.20) to yield the results shown in Table 6.

Table 6

Types of Broadening

Absorber	Disturber	n	Name of Effect
Hydrogen-like	Neutral	3	Resonance broadening
Hydrogen-like	Charged	2	Linear Stark effect
Not hydrogen-like	Neutral	6	Van der Waals broadening
Not hydrogen-like	Charged	4	Quadratic Stark effect

The case for n = 3 is also known as self broadening since it holds for any neutral particles disturbing an atom of the same kind. As stated above, this is important is astrophysics only for hydrogen as a general rule, although there are hydrogen deficient stars for which exceptions could be important. Note that all atoms become hydrogen-like if the outer electron is in a large enough state, so the Stark effect changes from quadratic to linear for high enough energy levels of any atom.

When the important interactions for a given line have been identified, their net effects on the line must be determined. The general case is extremely complicated, and usually only one of the limits of two extreme approximations is used.

These arise from consideration of the relative values of the collision time t_c and the radiation time t_r of the line. When the collision time is very short, the time dependent behavior of the atoms during a collision can be ignored; only the integrated effects of the collision as a whole need be considered. This approximation is known as the impact or phase shift theory. At the other extreme the radiation time is very short, many collisions may take place simultaneously, and the motions of the perturbers can be ignored. Here we must find the probability that the emitting atom finds itself in a static field of a given size. This is known as the statistical or static field approximation.

The collision time is of the order of d/v, where d is the impact parameter (closest distance of approach) of the collision and v is the relative velocity. The radiation time of a line is defined as $1/2\pi\Delta\nu_0$, where $\Delta\nu_0$ is the mean frequency shift from the line center. The ratio is then

$$\frac{t_r}{t_c} \simeq \frac{v}{2\pi\Delta\nu_0 d} \qquad (19.21)$$

High temperatures and small masses, which increase the relative velocity, tend to favor the impact theory; broad lines with large average shifts favor statistical theory.

We can actually in effect define a radiation time for each photon absorbed and emitted, with $t_r = 1/2\pi\Delta\nu$, and now $\Delta\nu$ is the shift of a given photon. We can then evaluate t_r/t_c for each photon and find the frequency shift for which $t_r = t_c$. On using equation (19.18) to eliminate d, we find

$$\Delta\nu_1 \simeq \left(\frac{v}{2\pi C^{1/n}}\right)^{n/n-1} \qquad (19.22)$$

For shifts very much greater than $\Delta\nu_1$, the static theory is applicable; for those much smaller than this, the impact theory is valid. For greater detail on impact vs statistical theory for different types of broadening, see the Jefferies, Mihalas, and Unsöld books listed in the general references.

For purposes of illustration simple examples of the limiting theories will now be given. For statistical theories the simplest example is the nearest neighbor approximation, in which it is assumed that only the nearest perturber to the atom has to be taken into account. Let there be N perturbing particles per unit volume; then $4\pi N r^2 dr$ is the probability of finding such a particle between r and r + dr of the origin. If $p^*(r)$ is the probability that no particle is within r, then

$$p^*(r + dr) = p^*(r)(1 - 4\pi N r^2 dr)$$

This can be integrated to yield

$$p^*(r) = e^{-(r/r_0)^3} \qquad (19.23)$$

where r_0 is the mean distance between particles:

$$r_0 = (\tfrac{4}{3}\pi N)^{-1/3} \qquad (19.24)$$

The probability $p(r)dr$ of finding the nearest particle between r and r + dr is then

$$p(r)dr = p^*(r)4\pi N r^2 dr = 3\frac{r^2}{r_0^2} e^{-(r/r_0)^3} \frac{dr}{r_0} \qquad (19.25)$$

A perturbing particle at r produces a frequency shift given by equation (19.18). If r is eliminated through this relation, we find the broadening function:

$$\phi_y dy = \frac{3}{n} e^{-(y^{-3/n})} y^{-(n+3)/n} dy \qquad (19.26)$$

where
$$y = \frac{\Delta\nu}{\Delta\nu_0} \qquad (19.27)$$

and
$$\Delta\nu_0 = \frac{C}{r_0^n} \qquad (19.28)$$

Note that y is always positive. This means that all shifts

are in the same direction, either to larger or to smaller frequencies, depending on the sign of the interaction constant C. The broadening is not symmetric about the line center. If the energy levels are degenerate, each separate line component has its own value of C, and the net line broadening must be calculated component by component.

The nearest neighbor approximation is valid only for very close encounters, so equation (19.28) holds in practice only for large values of y. In this region it is seen that ϕ_ν is proportional to $(\Delta\nu)^{-(n+3)/n}$. If all of the disturbing atoms are taken into account instead of only the nearest, the generalization of the above analysis leads to what is known as the Holtsmark distribution. For this derivation see the above mentioned references or S. Chandrasekhar, Rev. Mod. Phys. 15, 1, 1943.

Perhaps the simplest example of an impact type theory is one in which the atom is pictured as indefinitely radiating a wave of fixed frequency, but in which the wave is intermittantly interrupted by collisions. This obviously satisfies the requirement that the radiating time be much longer than the collision time.

If each collision completely disrupts the wave train, the train is a series of independent waves of duration equal to the free time between collisions. Let t be the time between collisions; then the wave is of the form $e^{-i\omega_0 t}$ for all times between zero and t. The Fourier transform of the wave is

$$F(\omega,t) = \frac{e^{i(\omega-\omega_0)t} - 1}{\sqrt{2\pi} i(\omega - \omega_0)} \quad (19.29)$$

The energy carried by waves of frequency ω is proportional to the square of the transform $|F(\omega,t)|^2$. The net energy carried by waves of this frequency is obtained by multiplying this by the probability of free time t, and then integrating over all times.

The probability of free times between t and t + dt is identical to the similar probability of mean free paths given by equation (2.4). If T is the mean free time between collisions, then the probability distribution is

$$p(t)dt = e^{-t/T} \, dt/T \quad (19.30)$$

Carrying out the indicated integration, we find

$$|F(\omega)|^2 = \int_0^\infty p(t)|F(\omega,t)|^2 dt = \frac{1}{2\pi} \frac{1}{(\omega - \omega_0)^2 + 1/T^2} \quad (19.31)$$

The broadening function is found by normalizing this quantity. If we introduce the collisional damping constant

$$\delta_c = \frac{1}{2\pi T} \quad (19.32)$$

and introduce the standard frequency through $\omega = 2\pi\nu$, we obtain the following broadening function:

$$\phi_\nu d\nu = \frac{\delta_c}{\pi} \frac{d\nu}{(\nu - \nu_0)^2 + \delta_c^2} \quad (19.33)$$

This model for impact broadening was considered by H. A. Lorentz in the early part of the present century. The function (19.33), which is of the same form as equation (19.16) for natural broadening, is often called the Lorentz profile. A more general derivation by E. Lindholm indicated that collisions cause a shift in the position of the line center as well as broadening the line. More recent quantum mechanical calculations show that non-adiabatic effects, collisionally induced transitions in the radiating atom, can be quite important in some circumstances.

A given line is broadened by several effects simultaneously, and the separate effects must be combined. A common example is one in which the pressure broadening is of the Lorentz type. In this case the pressure and natural effects are both of the Lorentz profile, and they will together be represented by equation (19.16), with δ being the sum of the natural and collisional damping constants. Of course, the collisional damping constant may itself be the sum of those of separate collisional mechanisms, as long as each of them can be adequately represented by the Lorentz profile.

The final profile is found by folding together the Lorentz function and the Doppler function (19.6). Consider an atom with instantaneous radial velocity v. It absorbs and emits according to the Lorentz profile, but the frequency that is observed will be Doppler shifted according to equation (19.1):

$$\phi_\nu(v)d\nu = \frac{\delta}{\pi} \frac{d\nu}{(\nu - \nu_0 v/c - \nu_0)^2 + \delta^2} \qquad (19.34)$$

Multiply this by the probability of radial velocity v, equation (19.3), and integrate over all v:

$$\phi_\nu d\nu = \phi_u du = \frac{a\,du}{\pi^{3/2}} \int_{-\infty}^{\infty} \frac{e^{-y^2}\,dy}{(u+y)^2 + a^2} \qquad (19.35)$$

The substitution $y = v/v_0$ is used. u is the frequency shift in units of the Doppler width D as in equation (19.7), and a is the damping constant in units of the Doppler width:

$$a = \frac{\delta}{D} \qquad (19.36)$$

The form of equation (19.35) is known as the Voigt profile. It is common to see the function $H(a,u) = \sqrt{\pi}\phi_u$ in the literature. The H function is not normalized to unity for integration over frequency (or u), but it does equal unity at the line center in the limit of small damping. The Voigt profile is tabulated in a number of places, for example, G. D. Finn & D. Mugglestone, M.N. 129, 221, 1965; and D. G. Hummer, Mem. R.A.S. 71, 271, 1965.

As will be seen in the next section, the central regions of a line, within about three Doppler widths of the center, are dominated by Doppler broadening. Beyond this Doppler core natural and pressure broadening are important. For small frequency shifts and for weak lines, therefore, it is not necessary to know the pressure broadening with high accuracy.

20. Profiles, Equilavent Widths, Curves of Growth

The various quantities needed to solve the line problem have been discussed in previous sections. It is now a matter of putting the pieces together. The intensity emitted in the normal direction by a plane atmosphere is

$$I_\nu = \int_0^\infty S_\nu(\tau) e^{-\tau} d\tau \qquad (20.1)$$

The equations can be written for other geometries and other directions of emergence, but these are unnecessary complications in the present context. Equation (20.1) is valid for both line and continuum radiation.

One generally starts the line problem with what one hopes is an accurate model atmosphere. By the methods outlines in Chapter 3, a table of physical conditions has been obtained as a function of τ_c, the continuum optical depth at a frequency close to that of the line being studied. If the line is assumed to be formed in LTE, the line source function is immediately known at each point in the atmosphere; however the line absorption needs to be found before equation (20.1) can be solved for the line radiation.

The atomic absorption coefficient for the line is given by equation (17.4), so it is necessary to calculate the broadening function ϕ_ν at each point in the atmosphere. The correction for induced emission must also be determined if it is important. For an assumed abundance of the element in question, the line absorption coefficient k_l is calculated, and the line and continuum coefficients added to obtain the total absorption coefficient $k = k_l + k_c$. In terms of this k one can integrate through the model atmosphere to find τ, the total optical depth. This makes the physical conditions known as functions of τ, and equation (20.1) can then be numerically integrated to obtain the theoretical line intensity. When this process has been repeated for a number of frequencies across the line, a complete picture of the line shape or profile emerges.

If LTE is not a satisfactory approximation for the line, the solution becomes much more difficult. The source function is given by equation (18.10) or a suitable modification if the two level relation is not sufficient. The source function depends upon the radiation which is being sought, so an iterative scheme might be necessary. If the lower level of the line has its population strongly influenced by transitions to other levels, the simultaneous solution for these several transitions must be made. The absorption coefficient as well as the source function is influenced by non-LTE effects, as it depends upon excitation and ionization conditions.

The calculated line profile depends upon many things: the model atmosphere, the broadening function, the mechanism of line formation, the assumed excitation and ionization conditions, and the abundance of the element. It is also necessary that the various atomic constants of the line to be known, such as the oscillator strength, damping constants, and possibly other radiative and collisional transitional probabilities. A comparison of observed and calculated line profiles provides a very complicated test of the accuracy with which the assumed values and theories represent actual conditions in a star. The observational checks can be made more meaningful by using many lines and, in the case of the Sun, by using the same line at different positions on the solar disk. Of course, the quality of the observations is a very important part of the test.

For many years it was not possible to obtain very accurate observations of line profiles, and most checks were made with the total strength of a line rather than with the detailed profile. While this observational limitation no longer exists, it is still much easier to measure the line strength than the profile. Further, the line strength is much less sensitive to some of the uncertainties which enter the line formation theory; if these uncertainties are not the direct objects of interest in an investigation, therefore, it would be advantageous to use only the total strengths of the lines.

The strength of a line is measured by its equivalent width (ew):

$$W = \int_0^\infty \frac{I_c - I_\nu}{I_c} d\nu \qquad (20.2)$$

I_c is the continuum intensity in the region of the line, and I_ν^c is the intensity in the line; both are monochromatic quantities. The ew W is measured in frequency units, but it is not a monochromatic quantity. Wavelength units can also be used. W can also be measured in intensity, as above, or in terms of the flux. Note that W is not a measure of the absolute strength of a line, but of its strength relative to that of the background continuum. The integrand of equation (20.2) is known as the residual intensity r:

$$r = \frac{I_c - I_\nu}{I_c} \qquad (20.3)$$

To illustrate how ew's are less sensitive to certain quantities than are profiles, it will now be shown than an ew does not depend upon macroturbulent velocities in the source. As indicated in the last section, macroturbulence is the term given to motions which have a scale much larger than the distance corresponding to optical depth unity. It is a property of fluxes, not intensities, so that flux ew's are used in this illustration.

The emitting surface of a source is divided into n separate regions indicated by the subscript i, i = 1,...,n. Each region is to be large enough to be optically thick, yet small enough to have an essentially constant turbulent velocity. The definition of macroturbulence makes this construction possible. Since the regions are optically thick, they act as independent sources and the total flux is the sum of the individual ones:

$$F_c = \sum_i F_c(i) \qquad F_\nu = \sum_i F_\nu(i) \qquad (20.4)$$

The residual flux in the total radiation is

$$r = 1 - \frac{F_\nu}{F_c} = 1 - \frac{\sum F_\nu(i)}{F_c} \qquad (20.5)$$

In terms of quantities for individual regions, $F_\nu(i) = F_c(i) - F_c(i)r(i)$. Equation (20.5) then gives

$$r = \frac{\sum F_c(i)r(i)}{F_c} \qquad (20.6)$$

As far as the line is concerned, F_c and other continuum quantities do not depend upon frequency; therefore, the total ew is

$$W = \frac{1}{F_c} \Sigma F_c(i) \int_0^\infty r(i) d\nu = \frac{1}{F_c} \Sigma F_c(i) W(i) \qquad (20.7)$$

The ew of the total flux is the average of the individual ew's weighted by the individual fluxes. Each $W(i)$ is independent of the macroturbulent velocities, and the total W must be also. The line profile is distorted by this velocity field, but the ew is not affected. Stellar rotation produces a velocity dispersion among the emitting surface elements that has the same properties as macroturbulence; rapid rotation of a star makes its lines broader and more shallow, but the ew's are unchanged. Microturbulence is quite different. Regions small enough to have a uniform microturbulent velocity are optically thin; they partially shield each other, and their fluxes are not simply additive.

It is not worthwhile to examine here in detail the numerical techniques that are used to calculate line profiles and ew's. An understanding of the basic physics for the nonspecialist is the aim of the present work, and here I shall only summarize some of the more basic properties of lines in stellar atmospheres. An oversimplified model of line formation will be used for illustration in order to keep the main physical principles from being hidden by the mathematical details. The so-called Milne-Eddington model for line formation and LTE are assumed. An alternate mode of line formation known as the Schuster-Schwarzschild approximation is discussed in some of the literature listed in the general references.

The first assumption of the Milne-Eddington model is that the continuum source function is a linear function of the continuum optical depth:

$$B_\nu(T) = B_0(b\tau_c + 1) \qquad (20.8)$$

Since LTE is assumed, $B_\nu(T)$ is the source function in both the continuum and the line. While there is no strong theoretical justification for the form of equation (20.8), it is an interpolation formula which contains one of the essential properties of a stellar atmosphere, namely, a moderate temperature increase with depth. Its simple form makes it easy to use from a mathematical viewpoint. B_o and b are adjustable constants which can be used to maximize the fit. Features which depend upon the reversal of the temperature gradient in the outer atmosphere obviously cannot be reproduced by this model.

The second major assumption of the Milne-Eddington approximation is that the ratio of line to continuum absorption does not depend upon depth:

$$\eta = \frac{k_l}{k_c} \neq f(\tau) \qquad (20.9)$$

It follows from this that $\tau_l = \eta \tau_c$, and the total optical depth is

$$\tau = \tau_l + \tau_c = (1 + \eta)\tau_c \qquad (20.10)$$

The two basic assumptions of the Milne-Eddington approximation are both artificial and are made only for mathematical convenience: they allow the line problem to be easily solved without doing too much violence to the physical principles involved. In fact equation (20.8) is fairly accurate with the proper choice of the constants for the appropriate part of the spectrum; equation (20.9) is very good for some lines and quite poor for others. Actually useful quantitative information can be obtained by comparing real stars with the Milne-Eddington model, but with modern computers a much more sophisticated theory can be used about as easily. The simple theory is justified today only because of its teaching value.

If equations (20.8) and (20.10) are substituted into equation (20.1), we find

$$I_\nu = B_o \left(\frac{b}{1 + \eta} + 1 \right)$$

$$I_c = B_o (b + 1) \qquad (20.11)$$

The residual intensity is

$$r = \frac{b}{1 + b} \frac{\eta}{1 + \eta} \qquad (20.12)$$

The line absorption affects the radiation only through η. For very weak absorption, $\eta \to 0$ and $I_\nu \to I_c$. In this case the depth of the line below the surrounding continuum is proportional to η, as equation (20.12) indicates. For very strong line absorption, $\eta \to \infty$ and $I_\nu \to B_0$, the surface value of the source function. In this case the residual intensity becomes independent of η and depends only upon b, which fixes the rate of increase of the source function with depth. For an isothermal atmosphere, b = 0 and $I_\nu = I_c$. If the source function is everywhere the same, it makes no difference whether the radiation originates in deep or in shallow layers.

In order to be more specific about line properties, the dependence of η with frequency across the line needs to be known. We let

$$\eta = \eta_0 f(u) \qquad (20.13)$$

where as before u is the distance from the line center in units of the Doppler width D, and η_0 is the value of η at the line center. Thus f(0) = 1, and f is proportional to the broadening function ϕ_u.

The Voigt profile is appropriate for many lines, and it will be assumed valid here; however, in keeping with the other simplifications made here, an approximate form will be used.

In most astrophysical applications, the damping constant a in equations (19.35) and (19.36) is small, usually in the range 0.001 - 0.1. Now consider equation (19.35) for u being moderately small. The denominator of the integrand has a sharp minimum at y = -u, and most of the contributions to the integral come from y values near -u. If a were not small, a larger range of y would be important. Over this limited range in y, e^{-y^2} is nearly constant and can be taken out from under the integral without loss of much accuracy. The remaining integral has the value π/a; therefore,

$$\phi_u du \simeq \frac{1}{\sqrt{\pi}} e^{-u^2} du \qquad (20.14)$$

This is exactly the form of the profile function for thermal Doppler broadening alone, as indicated by equation (19.6); therefore, near the line center Doppler broadening dominates over the effects of pressure and natural broadening.

Consider now equation (19.35) for very large values of u, which means far from the line center. The denominator still has a sharp minimum at $y = -u$, but now the factor e^{-y^2} is so small there that that region cannot contribute appreciably to the integral. Instead, the main range of y which is important to the integral is that which is centered on $y = 0$, i.e., values for which $y^2 \ll u^2$. The denominator is thus approximately equal to simply u^2, and we have

$$\phi_u du \simeq \frac{a du}{\pi^{3/2} u^2} \int_{-\infty}^{\infty} e^{-y^2} dy = \frac{a du}{\pi u^2} \qquad (20.15)$$

We see that this is a pure Lorentz profile for u much greater than the damping constant a, as a comparison with equation (19.33) indicates. At large distances from the line center natural and pressure broadening dominate over Doppler broadening.

The transition region between the extremes of equations (20.14) and (20.15) can be roughly found by equating the two expressions. For the appropriate values of the damping constant, the transition occurs at approximately $u = 3$, and this is insensitive to the precise value of a. The part of a line with $u \lesssim 3$ is known as the Doppler core of the line; beyond are the damping wings. Only very strong lines have wings: the absorption at $u = 3$ is about $e^{-9} = 10^{-4}$ times that at the line center, so a line must be very strong if there is to be appreciable absorption extending beyond $u = 3$.

Equation (20.13) can be written approximately as

$$\begin{aligned} \eta &= \eta_0 e^{-u^2} & u \lesssim 3 \\ \eta &= \frac{a \eta_0}{\sqrt{\pi} u^2} & u \gtrsim 3 \end{aligned} \qquad (20.16)$$

The approximate line profile is found by substituting into equation (20.11).

It is convenient for purposes of classification to divide lines into three classes: weak, medium, and strong. Weak lines are those for which the absorption is weak even in the line center, i.e., $\eta_0 \ll 1$. Medium strength lines have strong absorption at the center, but they become weak within the Doppler core. We noted above that the absorption at the edge of the core is about 10^{-4} of that at the center, so medium lines have $1 \ll n_0 \lesssim 10^4$. Strong lines are those with well developed wings, $n_0 \gtrsim 10^4$.

Equation (20.12) leads to the following for the residual intensity at the center of a line:

$$r_0 = \frac{b}{1+b} n_0 \quad \text{(weak)}$$

$$r_0 = \frac{b}{1+b} \quad \text{(medium \& strong)}$$

(20.17)

The central depth is proportional to the central absorption for weak lines, but it is independent of n_0 for medium and strong lines. This result is expected: when n_0 has reached a certain large value, the radiation from the line center comes essentially from the surface of the star. Increasing the line absorption further will not affect the central radiation, as it will still come from the surface. If, instead of equation (20.8), we had a temperature increase in the outer atmosphere, simulating the chromosphere of the Sun, the result would be quite different. Weak lines would not "see" the temperature rise because it would be transparent to their radiation; however, the centers of strong lines would see such regions. If the lines were formed in LTE, they would show emission peaks in their cores. If non-LTE effects are dominant, the source function is not particularly sensitive to the temperature, and an emission core may not occur even in lines which are strong enough to see the temperature rise. The understanding of which lines should show these emission cores and which should not was one of the early triumphs of the non-LTE theory.

The above also illustrates a more general point. From Equation (20.1) we see that if the source function is constant, $I_\nu = S_\nu$. If S_ν does vary with depth, the emergent intensity is an average of the source function over the depths which contribute to this intensity. Since the average depth of formation is one mean free path, we expect the intensity coming out of a star to be rather close numerically to the value of the source function one mean free path below the surface. This is a good rule of thumb which gives useful results when only rough calculations are sufficient.

We found in Section 8 that this rule of thumb is exact

if the source function is a linear function of optical depth. The Milne-Eddington source function (20.8) is linear, and the intensities in equations (20.11) are seen to be equal to the source function at a depth where the appropriate optical depth is unity. The residual intensities in equations (20.17) can also be understood from this argument.

A measure of the width of a line is given by u_2, the value of u for which the residual intensity falls to one-half of its central value. Equations (20.12) and (20.13) lead to the following:

$$\eta(u_2) = \frac{n_o}{2 + n_o} \qquad (20.18)$$

For the three classes of lines this reduces to

$$u_2 = 1 \qquad \text{(weak)}$$

$$u_2 = \sqrt{\ln n_o} \qquad \text{(medium)} \qquad (20.19)$$

$$u_2 = \sqrt{a n_o} \qquad \text{(strong)}$$

The line widths behave nearly opposite to the residual intensities: the widths of weak lines are independent of n_o, medium line widths are very insensitive to it, and strong line widths are proportional to the square root of the central absorption. Again, a rough interpretation is not difficult to understand. Weak and medium lines are broadened by the Doppler effect alone. Only an extremely small percent of the atoms have radial velocities greater than two or three times the average, which means that only this small percent can absorb at frequencies having u greater than 2 or 3. There are so few atoms having extreme radial velocities that increasing n_o does not appreciably increase the absorption beyond $u = 3$, and the line width is nearly independent of n_o. When damping becomes dominant, however, an atom does not need a large velocity in order to absorb at large u, and wide wings grow on the line as n_o increases.

Equations (20.17) and (20.19) indicate how the shapes of lines vary with central absorption, and this is illustrated

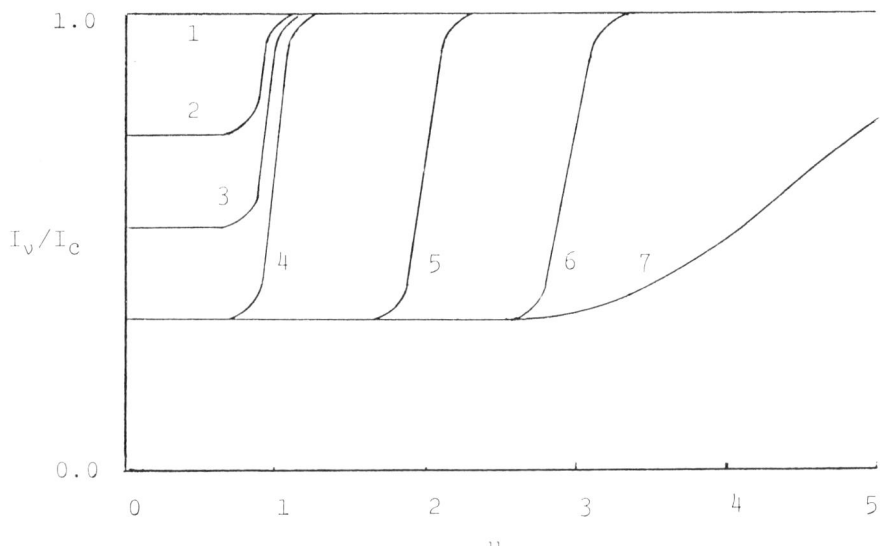

Figure 6. Weak, medium, and strong line profiles. See the text for a description.

in Figure 6. This figure is a plot of intensity, in units of the continuum intensity, versus u. Only half of the line profiles are shown, as they are symmetric about u = 0. Curve 1 is the straight line across the top and represents the continuum intensity. The seven curves differ from each other only through the value of η_0, the central absorption.

Curve 1 has η_0 = 0, i.e., no line at all. Curves 2 and 3 represent weak lines corresponding to small but increasing values of η_0. Note the different depths but the nearly constant widths for which u_2 is about unity. In line 4, the depth has reached saturation, but the width is still about out to u = 1. Line 4 is approaching medium in strength. Lines 5 and 6 are still medium in strength, being within the Doppler core of u ≃ 3; line 7 is a strong line with well-developed wings extending beyond the Doppler core.

It should be emphasized that Figure 6 does not represent absorption lines directly as they would be observed, for the conversion between u and frequency depends on temperature, as indicated in equations (19.7) and (19.8). Temperature varies with depth in the atmosphere of a star, so a fixed value of frequency does not correspond to a fixed value of u. This effect is rather small, however, so Figure 6 comes close to showing observed profiles, apart from the physical and mathematical approximations that were made.

The ew, measured in frequency units, is found by integrating the residual intensity of a line over frequency. In an approximate way this integral can be replaced by the product of r_0 and $\Delta\nu$, an average width of the line. In dimensionless units the line width is about $2u_2$, so $\Delta\nu \simeq 2u_2 D$. We then find from equations (20.17) and (20.19)

$$W = \frac{2bD}{1 + b} \eta_0 \qquad \text{(weak)}$$

$$W = \frac{2bD}{1 + b} \sqrt{\ln \eta_0} \qquad \text{(medium)} \qquad (20.20)$$

$$W = \frac{2bD\sqrt{a}}{1 + b} \sqrt{\eta_0} \qquad \text{(strong)}$$

One can use either frequency or wavelength units for the Doppler width D, and then the ew W will be in the same units. Remember that, among other things, η_0 is proportional to the abundance of the element producing the absorption line.

The relation between W and η_0 is known as the curve of growth. The curve divides itself into three regions. In the weak line region, W is directly proportional to η_0, i.e., to the abundance of the absorbing atoms. In the medium line region the curve saturates, and W is nearly independent of η_0, and in the strong line region W goes as the square root of η_0.

A rather large number of ways have been developed to use the curve of growth in one form or another to determine the abundances of the elements producing the spectral lines. The advent of high speed computers has made these methods mainly of historical interest only, as it is more accurate and just about at easy to use more sophisticated means. The interested reader can consult the general references listed in the front of the book.

    The analysis described in this section is very rough, but it does illustrate the more important physical principles of line formation. I believe that this is a good way to introduce the subject, particularly to those who want a quick summary of the material. Those who intend to become scientifically active in this field must, of course, go into many of the details which have been omitted here.

Chapter 5: Polarization

21. Pure Polarized Radiation

The main respect in which the analysis of Chapter 1 is incomplete is that the polarization state of radiation is not included. The importance of polarized radiation in many parts of modern astrophysics makes worthwhile the inclusion of a chapter on the subject, even though polarization is not very important in the context of stellar atmospheres.

If radiation is polarized, certain of its properties at one instant are correlated with those at a later instant: those properties are "remembered" to a certain extent, the extent being a measure of the degree of polarization. We will consider this quantitatively in the next section; here we consider the simplest case of pure polarized radiation.

The plane wave solution of Maxwell's equations in a vacuum for the electric (or magnetic) field of a wave that is propagating in the z direction is

$$E_x = a_x \sin(\omega t - \delta_x)$$
$$E_y = a_y \sin(\omega t - \delta_y)$$
(21.1)

Here ω is the frequency of the radiation ($\omega = 2\pi\nu$; because of the profusion of subscripts, the frequency subscript of quantities which are per unit frequency interval will no longer

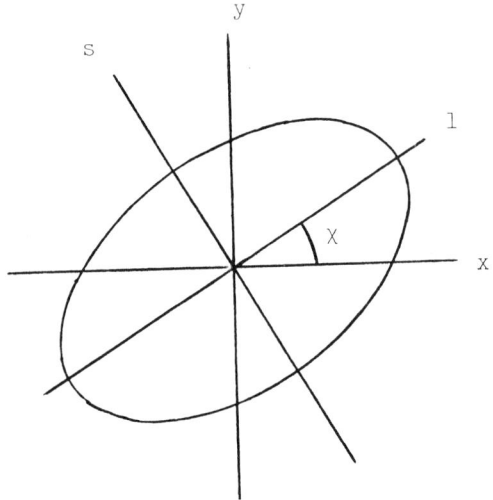

Figure 6. The polarization ellipse.

be expressed.), the a's are amplitudes and the δ's are phase constants. This is the parametric form of an ellipse, the so-called polarization ellipse of Figure 6. The electric vector makes one circuit of the ellipse in one period of the radiation.

The intensity of this radiation being propagated in the z direction is proportional to the square of the amplitude; we will assume for simplicity that the proportionality factor is already included in the above equations, so we have directly

$$I = a_x^2 + a_y^2 = I_x + I_y \qquad (21.2)$$

I_x and I_y should not be thought of as components of the intensity, as it is a scalar quantity; rather they are the partial intensities due to vibrations in the given directions.

Figure 6 also shows the coordinate system with axes along the principal axes of the ellipse: the l axis is along the major axis and the s axis is along the minor axis of the polarization ellipse. In this system the equations (21.1) are

$$E_l = a_l \sin(\omega t - \delta_1)$$
$$E_s = \pm a_s \cos(\omega t - \delta_1) \quad (21.3)$$

In the principal axes system the phase difference is $\pm\pi/2$; this is reflected in the $\pm$ sign in front of a_s. The different signs on E_s correspond to different senses of rotation of the electric and magnetic vectors: plus is right handed rotation and minus is left handed.

Instead of the amplitudes as given in equations (21.3), we can introduce the new quantities a_o and β as follows:

$$a_l = a_o \cos \beta \qquad \pm a_s = a_o \sin \beta \quad (21.4)$$

or equivalently,

$$a_o^2 = a_l^2 + a_s^2 \qquad \tan \beta = \frac{\pm a_s}{a_l} \quad (21.5)$$

The angle β is between $-45°$ and $+45°$; its sign determines the handedness of the sense of rotation, and its magnitude fixes the ellipticity of the polarization ellipse. For $\beta = \pm 45°$, the ellipse is a circle; for $\beta = 0$, the ellipse degenerates into a straight line.

The relation between the wave components in the xy and in the ls systems is given by the transformation of a rotation through the angle χ:

$$E_x = E_l \cos \chi - E_s \sin \chi$$
$$E_y = E_l \sin \chi + E_s \cos \chi \quad (21.6)$$

Now if equations (21.3) and (21.4) are substituted into (21.6), we find expressions for E_x and E_y; Equations (21.1) are alternate expressions for E_x and E_y. These two sets of expressions must, of course, be equivalent. If we expand all of these equations so each term has either a sin ωt or a cos ωt, the coefficients of these functions must separately vanish. This leads to the following four relations:

$$a_x \cos \delta_x = a_0(\cos \beta \cos \chi \cos \delta_1 \pm \sin \beta \sin \chi \sin \delta_1)$$

$$a_x \sin \delta_x = a_0(\cos \beta \cos \chi \sin \delta_1 \mp \sin \beta \sin \chi \cos \delta_1)$$

$$a_y \cos \delta_y = a_0(\cos \beta \sin \chi \cos \delta_1 \mp \sin \beta \cos \chi \sin \delta_1)$$

$$a_y \sin \delta_y = a_0(\cos \beta \sin \chi \sin \delta_1 \pm \sin \beta \cos \chi \cos \delta_1)$$

(21.7)

A bit of algebra applied to these equations then leads to the following four identities:

$$a_x^2 + a_y^2 = a_0^2$$

$$a_x^2 - a_y^2 = a_0^2 \cos 2\chi \cos 2\beta$$

$$2a_x a_y \cos(\delta_x - \delta_y) = a_0^2 \sin 2\chi \cos 2\beta$$

$$2a_x a_y \sin(\delta_x - \delta_y) = a_0^2 \sin 2\beta$$

(21.8)

The relations (21.8) show how the amplitudes and phase constants in the xy system are related to the properties of the polarization ellipse. The first equation can be seen from equation (21.2) to be the intensity of the radiation, which is related to the size of the ellipse a_0. The other three relations of (21.8) also have the dimensions of intensities, although the quantities do not represent physical energy in transit. These quantities were introduced by G. G. Stokes in 1852 for treating polarized light. We can write

$$I = a_o^2$$

$$Q = I \cos 2\chi \cos 2\beta$$

$$U = I \sin 2\chi \cos 2\beta \qquad (21.9)$$

$$= Q \tan 2\chi$$

$$V = I \sin 2\beta$$

The four quantities I, Q, U, and V are known as the Stokes parameters of pure polarized radiation. We see that the size of the polarization ellipse involves only I; the orientation of the ellipse involves Q and U; and the ellipticity brings in V.

Note that for linear or plane polarization, $\beta = 0$ and $V = 0$. For circular polarization, $\beta = \pm\pi/4$ and $Q = U = 0$. From this we see that Q and U tell us about the linear component of the polarization, and V indicates the circular part. I and V are intrinsic properties of the radiation, but both Q and U depend upon the coordinate system used to describe the radiation. We also see that

$$I^2 = Q^2 + U^2 + V^2 \qquad (21.10)$$

Under conditions described here, therefore, the four Stokes parameters are not independent of each other.

22. General Polarized Radiation

Although equations (21.1) do satisfy Maxwell's equations, the real situation is usually much more complicated than these equations suggest. In practice the amplitudes and phases are very rapidly fluctuating in value; if these fluctuations are completely incoherent or random, the equations describe all points within a circle whose radius depends on the maximum size of the amplitude. Any measurement of the properties of the radiation are necessarily averaged over many such fluctuations; the question is: How are we to interpret the polarization ellipse under these conditions?

With changing values of the amplitudes and phases, β and χ are no longer well-defined quantities, and equations (21.9) are no longer appropriate as they stand. We can, however, still use the quantities that appear in equations (21.8) if we indicate the appropriate time averages. Thus, we can have

$$I = \overline{a_x^2} + \overline{a_y^2}$$

$$Q = \overline{a_x^2} - \overline{a_y^2}$$

$$U = \overline{2a_x a_y \cos(\delta_x - \delta_y)} \tag{22.1}$$

$$V = \overline{2a_x a_y \sin(\delta_x - \delta_y)}$$

Suppose that the fluctuations in the amplitudes and phases are completely random; then it is apparent from the above that we would have $Q = U = V = 0$. This is the case of radiation which is completely unpolarized, and only the intensity I is needed to completely specify its properties.

We have seen two opposite extremes: When the amplitudes and phases are constants, the polarization ellipse is unique and equations (21.9) and (21.10) hold; when the amplitudes and phases fluctuate randomly, there is no polarization ellipse and I is the only non-zero Stokes parameter. More generally we have a situation between these extremes. This occurs when the fluctuations are correlated to some degree, i.e., all values within certain ranges are possible, but some are more likely than others. In this case we can see from equations (22.1) that Q, U, and V may be non-zero, although the polarization ellipse along with β and χ are not yet well-defined.

In order to develop further the idea of partially polarized radiation, we now consider radiation which has been artifically adjusted in phase by an arbitrary amount. In equations (21.1), which we apply to a certain instant of time, we decrease the phase of E_x by the amount ε with respect to E_y. Noting that it is only the phase difference that is of importance, we also set $\delta_y = 0$ and omit the subscript on δ_x:

$$E_x = a_x \sin(\omega t - \delta - \varepsilon)$$
$$E_y = a_y \sin \omega t \qquad (22.2)$$

The field in a direction making the angle ϕ to the x axis is

$$E_\phi = E_x \cos \phi + E_y \sin \phi \qquad (22.3)$$

If this is expanded out in terms of $\sin \omega t$ and $\cos \omega t$, we have

$$E_\phi(\varepsilon) = \{a_x \cos \phi \cos(\delta + \varepsilon) + a_y \sin \phi\} \sin \omega t$$
$$- a_x \cos \phi \sin(\delta + \varepsilon) \cos \omega t \qquad (22.4)$$

The partial intensity $I(\phi,\varepsilon)$ in this direction is the time average of the square of this amplitude; the result in expanded form is

$$I(\phi,\varepsilon) = \overline{a_x^2} \cos^2 \phi + \overline{a_y^2} \sin^2 \phi +$$
$$+ \sin 2\phi (\cos \varepsilon \, \overline{a_x a_y \cos \delta} - \sin \varepsilon \, \overline{a_x a_y \sin \delta}) \qquad (22.5)$$

This time averaging is, of course, taken at constant ϕ and ε. If we now substitute the Stokes parameters as given in the equations (22.1), we finally obtain

$$I(\phi,\varepsilon) = \frac{1}{2}\{I + Q \cos 2\phi + (U \cos \varepsilon - V \sin \varepsilon) \sin 2\phi\} \qquad (22.6)$$

This shows that the four Stokes parameters or their equivalent are both necessary and sufficient to determine the partial intensity in an arbitrary direction in which an arbitrary phase shift has been introduced.

Suppose that we have two or more separate beams of radiation which are independent in the sense that they have no permanent phase relations between them. If the beams are combined, the intensities are then additive; it follows from equation (22.6) that all of the Stokes parameters are then additive. This result does not hold for beams that are not independent, i.e., beams that interfere with each other in some manner.

Let us use the symbol $\vec{I}$ to denote the vector intensity, i.e., the vector whose four components are the four Stokes parameters (I,Q,U,V). Then for independent beams we have the combining of $\vec{I}_1$ and $\vec{I}_2$ as a single beam $\vec{I}$:

$$\vec{I} = \vec{I}_1 + \vec{I}_2 \qquad (22.7)$$

which means that $I = I_1 + I_2$, $Q = Q_1 + Q_2$, etc.

It is apparent that a given beam of radiation $\vec{I}$ can be divided into equivalent component beams in an infinite number of ways; there is one such division that is of particular importance: one beam is unpolarized and the other is completely polarized. Moreover, this division is unique for a given beam.

Let $\vec{I}_1$ be the unpolarized beam and $\vec{I}_2$ the completely polarized one. Then we obviously have $Q_1 = U_1 = V_1 = 0$, the condition for the total lack of coherence in the fluctuations of amplitudes and phases. To satisfy equation (22.7), we must then have $Q_2 = Q$, $U_2 = U$, and $V_2 = V$. But the second beam is completely polarized, so it must satisfy equation (21.10). It follows that the two component beams are given by

$$\vec{I}_1 = \begin{bmatrix} I - \sqrt{Q^2+U^2+V^2} \\ 0 \\ 0 \\ 0 \end{bmatrix} \quad \text{(unpolarized)} \qquad (22.8)$$

$$\vec{I}_2 = \begin{bmatrix} \sqrt{Q^2+U^2+V^2} \\ Q \\ U \\ V \end{bmatrix} \quad \text{(pure polarized)} \qquad (22.9)$$

The degree of polarization of a general beam of radiation is determined by the relative amounts of unpolarized and pure polarized radiation in its component beams. We see from the above equations that this is given by

$$\Pi = \frac{\sqrt{Q^2+U^2+V^2}}{I} \qquad (22.10)$$

Since there is no physical meaning to a negative intensity, we see from equation (22.8) that I^2 is always greater than or equal to $Q^2+U^2+V^2$, and the equality occurs only when the radiation is completely polarized.

The polarization ellipse is well-defined only for pure polarized radiation; it has no meaning for unpolarized radiation. We can introduce this ellipse for general radiation by dividing the beam into the two components as above, where one is unpolarized and the other is completely polarized. Then the completely polarized component is used to define the polarization ellipse as in Section 21.

Equations (21.9) relate the Stokes parameters of pure polarized radiation to the angles β and χ; β determines the ellipticity of the polarization ellipse through equations (21.4) and (21.5), while χ gives the orientation of the long axis of the ellipse as shown in Figure 6. If we rewrite these equations for only the completely polarized component, we will have

$$\cos 2\chi \cos 2\beta = \frac{Q}{\sqrt{Q^2+U^2+V^2}}$$

$$\sin 2\chi \cos 2\beta = \frac{U}{\sqrt{Q^2+U^2+V^2}} \qquad (22.11)$$

$$\sin 2\beta = \frac{V}{\sqrt{Q^2+U^2+V^2}}$$

The four Stokes parameters of arbitrarily polarized radiation are then equivalent to the total intensity, the degree of polarization, and the angles β and χ of the polarization ellipse.

We again note that if $\beta = 0$, the polarization ellipse

is a straight line, V = 0, and the polarization is linear. For β = ±π/4, the polarization is circular and Q and U both vanish. The degrees of linear and of circular polarization can be defined in analogy with equation (22.10):

$$\Pi_L = \frac{\sqrt{Q^2+U^2}}{I}$$
$$\Pi_C = \frac{V}{I}$$
(22.12)

The equation of transfer (4.1) indicated how the intensity varies along a path of propagation due to the emission and absorption of matter along that path. How is this to be generalized for the case of polarized radiation?

Polarized radiation has four "intensities" that are needed to characterize it, so we can expect four equations of transfer. Specialized emission and absorption coefficients may be needed for each Stokes parameter; furthermore, the propagation of each parameter may depend upon the values of the other parameters, which means that the four equations of transfer may be coupled together. This suggests the following form for the generalized equation of transfer:

$$\frac{d\vec{I}}{ds} = \vec{j} - \underline{k}\vec{I}$$
(22.13)

$\vec{I}$ is the vector intensity whose four components are the four Stokes parameters, and $\vec{j}$ has components which give the polarization properties of the emitted radiation. The first component, for example, is the same as the emission coefficient of equation (4.1). Note again that both $\vec{I}$ and $\vec{j}$ are monochromatic quantities. $\underline{k}$ is a 4x4 matrix which represents a generalized absorption coefficient, although not all of its component actually bring about a true absorption. This point will be noted in the next section.

As in the unpolarized case, further progress cannot be made in a given problem unless the emission and absorption processes are identified and can be calculated. In the next two sections applications to two astrophysically interesting

cases will be given.
   Much of this introductory material on polarization is based on the presentation of Chandrasekhar in Section 15 of his book Radiative Transfer.

23. Transfer in a Magnetic Plasma

   The absorption matrix $\underline{k}$ brings about the coupling of the four equations of transfer (22.13). Although $\underline{k}$ has 16 elements, most of these are usually small enough to be set to zero. The four diagonal elements k_{ii} are given by the ordinary absorption coefficient considered in the earlier parts of the book. This absorption is unpolarized in the sense that its effects do not depend upon the polarization state of the radiation. There are other terms in $\underline{k}$ which represent polarised absorption; and finally there are elements in $\underline{k}$ that bring about propagation effects that are not absorptions at all.
   A common situation is the propagation of radiation in a low density plasma with a magnetic field. The emission and transfer of synchrotron radiation in radio sources is an example. Whether the elements of $\underline{k}$ are due to a thermal or a relativistic plasma will not concern us at present. For an evaluation of the emission and absorption coefficients see, for example, the books by A. G. Pacholczyk, Radio Astrophysics, Freeman, 1970, and Radio Galaxies, Pergamon, 1977; V. L. Ginzburg, The Propagation of Electromagnetic Waves in Plasmas, Pergamon, 1964; or the two articles by V. L. Ginzburg and S. I. Syrovatskii in Ann. Rev. Astron. Astrophys., vol. 3 p. 297, and vol. 7, p. 375; and the references listed in the above.
   For the case we are considering, the elements k_{23} and k_{32} are quite important, and one is the negative of the other. We will also include the regular absorption elements k_{ii}, and the polarized terms k_{14} and k_{41}, which are equal. The other elements of $\underline{k}$ will be set to zero. To simplify notation, we set

$$k_{ii} = k, \quad i = 1,2,3,4$$

$$k_{23} = -k_{32} = f \qquad (23.1)$$

$$k_{14} = k_{41} = v$$

The equations of transfer written out in component form are

$$\frac{dI}{ds} = j_I - kI - vV$$

$$\frac{dQ}{ds} = j_Q - kQ - fU$$

$$\frac{dU}{ds} = j_U + fQ - kU$$

$$\frac{dV}{ds} = j_V - vI - kV$$

(23.2)

The equations separate into two sets of coupled relations; I and V are coupled and, independently, Q and U are interrelated. These equations illustrate the general property that $k_{ij} = \pm k_{ji}$, i.e., the coupling between two Stokes parameters is either symmetric or antisymmetric. The symmetric elements represent absorption, while the antisymmetric ones give rise to propagation or Faraday effects, as they are often called.

We now make the assumption of a uniform medium, so that all of the coefficients in equations (23.2) can be considered constants. This does not affect the basic physics that I wish to illustrate, and it greatly simplifies the mathematics.

Let us first look at the equations for I and V. These equations can be uncoupled by introducing new variables as follows:

$$I = I^e + I^o$$

$$V = I^e - I^o$$

(23.3)

The superscripts o and e stand for ordinary and extraordinary, the so-called normal modes of propagation in a medium. The equations for the new variables are

$$\frac{dI^e}{ds} = \frac{j_I + j_V}{2} - (k+v)I^e$$

$$= j_e - k_e I^e$$

$$\frac{dI^o}{ds} = \frac{j_I - j_V}{2} - (k-v)I^o$$

$$= j_o - k_o I^o$$

(23.4)

The ordinary and extraordinary beams are completely decoupled, and each has its own emission and absorption coefficients. Both j_V and v can be either positive or negative, depending on the direction of the magnetic field with respect to the direction of propagation, so the above equations in themselves do not specify which beam has the greater emission or absorption. These equations illustrate the meaning of the earlier statement that v represents polarized absorption.

The solutions of equations (23.4) are analogous to equation (4.7):

$$I^i = I_o^i e^{-k_i s} + \frac{j_i}{k_i}(1 - e^{-k_i s})$$

(23.5)

where the index i represents either o or e. I_o^i is the incident intensity at the boundary of the medium s = 0. One can now combine the two beams according to equations (23.3) to obtain I and V for the total radiation.

Suppose we arbitrarily suppress the ordinary beam by setting its incident intensity and emission coefficient to zero. Since only the extraordinary beam remains, we obviously have $I = I^e$; but from equation (23.3), we also have $V = I^e$, or $V = I$. According to equation (22.10), the extraordinary beam must by 100% polarized, and it must also have $Q = U = 0$ in order to avoid having Π greater than unity. In other words the extraordinary beam is pure circularly polarized radiation. If we had suppressed the extraordinary beam, we would have found that $V = -I$; the ordinary beam is also 100% circularly polarized, but with the opposite handedness of the extraordinary beam.

This illustrates the fact that, when radiation enters

a medium such as is described here, it separates into two beams which are each pure circularly polarized beams with the opposite handedness. The beams each have their own transmission properties, including indexes of refraction. The beams combine again on leaving the medium and entering a vacuum. According to equation (23.3), if $I^e = I^o$, then $V = 0$ for the total radiation. In this case the opposite handedness of the ordinary and extraordinary beams completely cancel each other leaving no circular component of the total radiation.

The solutions of equations (23.2) for Q and U do not conveniently separate into normal components, as the ordinary and extraordinary rays both have Q and U vanish. They do not vanish for the total radiation, however. This appears to violate the additive property of the Stokes parameters, but the normal rays are not independent beams. Their electric fields vibrate synchronously with a permanent phase difference, resulting in interference with each other. Under these conditions, the Stokes parameters are not additive.

We look for solutions for Q and U having the form of constants plus terms with exp-ks as a factor. The result is

$$Q = Q_p + \sqrt{A^2+B^2}\ e^{-ks}\ \cos(fs-\delta)$$
$$U = U_p + \sqrt{A^2+B^2}\ e^{-ks}\ \sin(fs-\delta) \quad (23.6)$$

where
$$Q_p = \frac{kj_Q - fj_U}{k^2 + f^2} \qquad U_p = \frac{fj_Q + kj_U}{k^2 + f^2}$$
$$A = -U_o + U_p \qquad B = Q_o - Q_p \qquad \tan \delta = \frac{A}{B} \quad (23.7)$$

We examine first the special case of external radiation being transmitted through the medium, which itself does not emit. Thus we set the emission coefficients to zero, and

$$Q = \sqrt{Q_o^2+U_o^2}\ e^{-ks}\ \cos(fs - \delta)$$
$$U = \sqrt{Q_o^2+U_o^2}\ e^{-ks}\ \sin(fs - \delta) \quad (23.8)$$

From the above and equations (22.11) we find that

$$\frac{U}{Q} = \tan(fs-\delta) = \tan 2\chi \qquad (23.9)$$

or

$$\chi = \frac{1}{2} fs + \text{const} \qquad (23.10)$$

The long axis of the polarization ellipse slowly rotates as the radiation propagates through the medium. This effect is known as Faraday rotation. If the properties of the medium are dominated by thermal electrons, then f is given by

$$f = \frac{e^3 N_e H \cos\theta}{\pi m^2 c^2 \nu^2} \qquad (23.11)$$

where N_e is the electron density, H is the magnetic field, θ is the angle between the propagation direction and the magnetic field, and ν is the frequency of the radiation. By observing at different frequencies, radio astronomers can sometimes identify Faraday rotation in the radiation of distant sources; this helps to determine some of the properties of the intervening plasma.

Finally we consider the case in which the radiation originates within the medium ($Q_0 = U_0 = 0$) and there is no significant absorption. Then equations (23.6) reduce to

$$\begin{aligned} Q &= -\frac{j_U}{f} + \frac{\sqrt{j_Q^2 + j_U^2}}{f} \cos(fs-\delta) \\ U &= \frac{j_Q}{f} + \frac{\sqrt{j_Q^2 + j_U^2}}{f} \sin(fs-\delta) \end{aligned} \qquad (23.12)$$

After a bit of algebra, we find that this leads to

$$f^2(Q^2+U^2) = 4(j_Q^2+j_U^2) \sin^2 fs/2 \qquad (23.13)$$

Under the assumed conditions, the total intensity is found from equations (23.3) and (23.5) to be simply $j_I s$, as in the unpolarized case. The degree of linear polarization is then

$$\Pi_L = \frac{\sqrt{Q^2+U^2}}{I} = \frac{\sqrt{j_Q^2+j_U^2}}{j_I} \left|\frac{\sin(fs/2)}{fs/2}\right| \qquad (23.14)$$

The degree of linear polarization is a sinusoidal function of (fs/2), limited to positive values only, with an amplitude that decreases as (fs/2). This interesting result is due to the combining of the propagating radiation with that which is newly emitted. With the assumption of a uniform medium, all radiation is emitted with the same orientation; however, the plane of polarization is subject to Faraday rotation as it is transmitted through the medium, and combining beams of different orientation will depolarize the radiation. If the product fs has such a value that the locally emitted radiation is in phase with that emitted near s = 0, reinforcement occurs and the degree of polarization has a local maximum. Noting again equation (23.11) for f, we see how observations at different frequencies can show this Faraday depolarization in nearly uniform radio sources of low optical thicknesses.

One can find much greater detailed discussions of the equation of transfer and its solutions for magnetic plasmas in the two books by A. G. Pacholczyk mentioned earlier in this section, plus the references given therein.

24. Rayleigh Scattering and the Sunlit Sky

In this section the transfer of polarized radiation will be applied to a scattering atmosphere being illuminated from the outside. Rayleigh scattering will be assumed, and the nature of this process is such that it will be convenient to use a modified set of Stokes parameters. Instead of I and Q, we introduce I_r and I_l, the partial intensities in two mutually perpendicular directions. The significance of the subscripts is that they are the final letters of the words perpendicular and parallel. The vector intensity then has the four components I_r, I_l, U, V. The parameters I and Q are then given simply by $I_r \pm I_l$.

In the equations of transfer (22.13), only the diagonal elements of $\underline{k}$ will be kept, the ones that represent unpolarized absorption. As before, we let $k_{ii} = k$, $i = 1,2,3,4$. The emission coefficients will be given by a suitable generalization of equations (5.8) and (5.10), as required for Rayleigh scattering.

Rayleigh scattering is best described in terms of the components perpendicular and parallel to the scattering plane, the plane that contains both the incident and the scattered rays. For these components, the phase relations are not changed by the scattering, while the amplitude of the parallel component is decreased with respect to the perpendicular one by the factor of $\cos \Psi$, where Ψ is the scattering angle. If we now examine equations (22.1), with x,y identified with r,l, we see that the scattered components of I_r, I_l, U, V will be proportional to $(1, \cos^2\Psi, \cos\Psi, \cos\Psi)$ respectively. We can then write for the vector emission coefficient

$$\vec{j} = k \int \underline{P}(\Psi) \vec{I}(\theta', \phi') \frac{d\omega'}{4\pi} \qquad (24.1)$$

where the phase matrix $\underline{P}$ is given by

$$\underline{P}(\Psi) = \frac{3}{2} \begin{pmatrix} 1 & 0 & 0 & 0 \\ 0 & \cos^2\Psi & 0 & 0 \\ 0 & 0 & \cos\Psi & 0 \\ 0 & 0 & 0 & \cos\Psi \end{pmatrix} \qquad (24.2)$$

Equations (5.8) and (5.10) are regained from the above if the incident radiation is unpolarized, so that $I_r = I_l = I/2$; note that $j_I = j_r + j_l$.

Radiation from the Sun is incident on the atmosphere from above. This radiation is assumed to be unpolarized and directed in a parallel beam along spherical angles of incidence (θ_o, ϕ_o). With $\mu_o = \cos \theta_o$, the total intensity of the incident radiation is given by

$$I_{in} = F\delta(\mu - \mu_o)\delta(\phi - \phi_o) \qquad (24.3)$$

where F is the flux of the sunlight and the δ's are Dirac δ-functions. The incident vector intensity, written horizontally to save space, is

$$\vec{I}_{in} = F\delta(\mu-\mu_o)\delta(\phi-\phi_o)(\tfrac{1}{2},\tfrac{1}{2},0,0) \qquad (24.4)$$

For convenience the angle θ is now measured positive in the downward direction toward the ground.

The scattering problem will not be treated exactly, but we will use single scattering theory. This means that photons that are scattered two or more times are ignorred, and the problem is very much simplified. The Earth's atmosphere is optically thin enough in visible light for this approximation to be reasonably good.

To be observed from the ground along angles θ,φ, a photon must enter the atmosphere along angles θ_o, ϕ_o; then at the correct point in the atmosphere it must be scattered through the angle Ψ so that it comes to the observer along the appropriate angles. The observed intensity is obtained by integrating the emission coefficient (24.1) over the observed line of sight; the quantity that goes into the integrand of (24.1) is the incident intensity (24.4), attenuated by the absorption prior to the scattering.

Let s' be the distance a photon travels between entering the atmosphere and being scattered; let s be the distance between the scattering point and the observer. Then it is apparent that

$$s'\mu_o + s\mu = z \qquad (24.5)$$

where z is the vertical height of the atmosphere. For a plane parallel atmosphere equation (24.5) can be immediately converted to the corresponding optical distances:

$$\tau'\mu_o + \tau\mu = \tau_o \qquad (24.6)$$

τ_o is the vertical optical depth of the atmosphere, which must not be too large if the single scatter theory is to give reliable results. Note that equation (24.6) does not rest upon the assumption that the atmosphere is uniform with height.✓

Following equation (4.6), we can write the integral for the once scattered skylight as

$$\vec{I} = \int_0^{z/\mu} \vec{j} e^{-\tau} ds = \int_0^{\tau_0/\mu} \frac{\vec{j}}{k} e^{-\tau} d\tau$$

$$= \int_0^{\tau_0/\mu} \{\int \underline{P}(\Psi)\vec{I}_{in} e^{-\tau'} \frac{d\omega'}{4\pi}\} e^{-\tau} d\tau \qquad (24.7)$$

Because of the delta-function nature of $\vec{I}_{in}$, the integral over $d\omega'$ simply changes functions of θ', ϕ' to functions of θ_0, ϕ_0. If we now introduce the vector $\vec{J}$ having the four components $(1, \cos^2\Psi, 0, 0)$, then we easily see that equation (24.7) integrates out to give

$$\vec{I} = \frac{3}{16\pi} F\vec{J} \frac{\mu_0}{\mu_0 - \mu} (e^{-\tau_0/\mu_0} - e^{-\tau_0/\mu}) \qquad (24.8)$$

For $\mu = \mu_0$, i.e., looking directly at the Sun, the last two factors of (24.8) reduce to $(\tau_0/\mu \, e^{-\tau_0/\mu})$.

Equation (24.8) is actually the solution to the problem. Writing out the components in full, we have

$$I_r = \frac{3Fg}{16\pi}$$

$$I_l = \frac{3Fg}{16\pi} \cos^2\Psi$$

$$U = 0$$

$$V = 0$$

(24.9)

where

$$g = \frac{\mu_0}{\mu_0 - \mu} (e^{-\tau_0/\mu_0} - e^{-\tau_0/\mu}) \qquad (24.10)$$

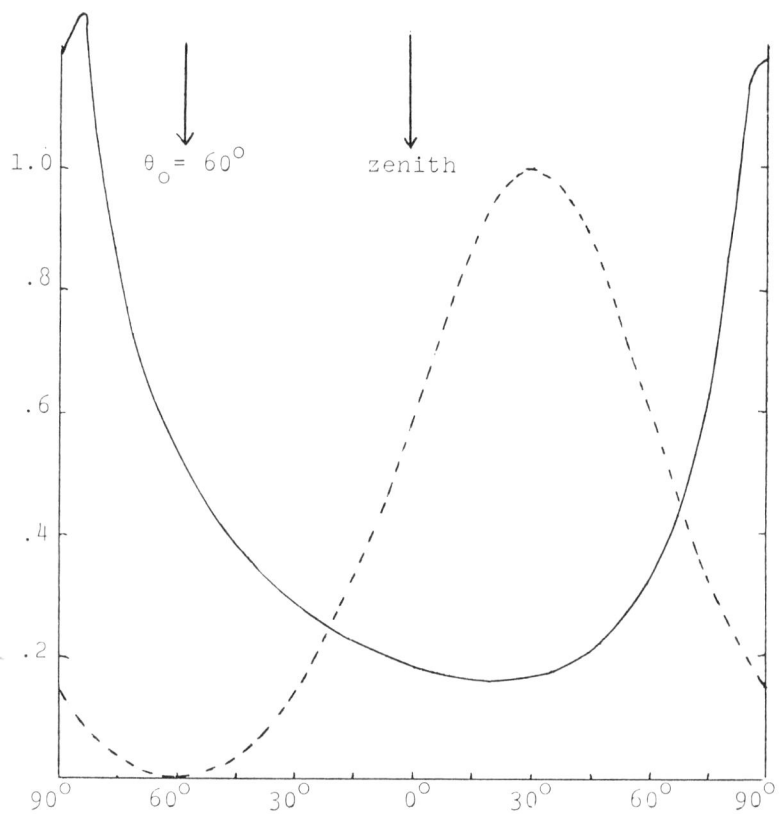

Figure 7. Intensity (solid curve) and degree of polarization (dotted curve) of the daylight sky for different zenith angles θ of observation. The intensity is in units of $3F/16\pi$. The atmosphere has an assumed normal optical thickness of $\tau_o = 0.2$, and the sunlight is incident at the angle $\theta_o = 60$ degrees. The observations are made along the meridian defined by the incident sunlight.

The total intensity of the skylight is

$$I = I_r + I_l = \frac{3Fg}{16\pi}(1 + \cos^2\Psi) \qquad (24.11)$$

There is no circular polarization, and the degree of linear polarization is given by

$$\Pi_L = \frac{I_r - I_l}{I_r + I_l} = \frac{1 - \cos^2\Psi}{1 + \cos^2\Psi} \qquad (24.12)$$

These quantities are illustrated in Figure 7 on the previous page for the case of $\theta_o = 60$ degrees and $\tau_o = 0.2$. The figure is prepared for the angle θ extending along the same meridian as that of the incident sunlight. The polarization is of course maximum at the angle of 90 degrees from the Sun, where the intensity is near minimum. Single scatter theory is not reliable near the horizon. Also, multiple scattering will add partially polarized radiation to that which is singly scattered, keeping the maximum polarization below 100%.

It must be remembered that equations (24.9) refer to a coordinate system fixed to the scattering plane, which includes the line of sight and the direction to the Sun. If a more convenient system is desired, i.e., one fixed to the meridian passing through the line of sight, then the components in equations (24.9) must be subjected to the appropriate rotation.

Figure 8 on the next page shows the scattering geometry in the atmosphere. The solid lines separated by the angle Ψ represent the incoming rays before and after the scattering. The angle ε is the amount one would have to rotate the scattering plane in order for it to coincide with the meridian. By applying the sine law of spherical trigonometry to the triangle that converges toward the zenith, we find

$$\sin \varepsilon = \frac{\sin \theta_o \sin(\phi_o - \phi)}{\sin \Psi} \qquad (24.13)$$

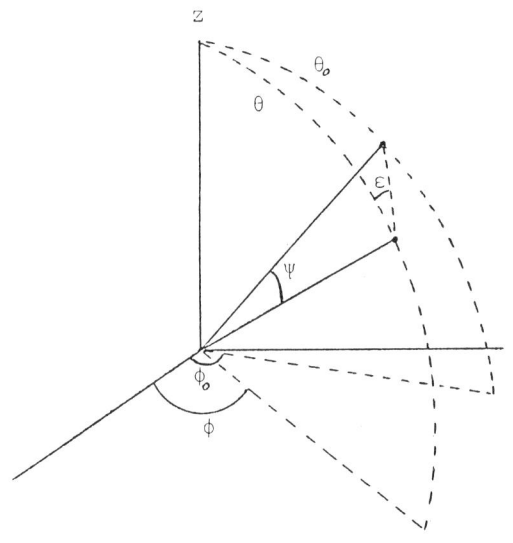

Figure 8. Scattering geometry

We must now determine how the rotation through the angle affects the Stokes parameters.

Referring back to Figure 6 on page 98, suppose that the xy system is rotated in the right hand fashion through the angle ε, which means that the orientation angle effectively becomes $(\chi - \varepsilon)$. The total intensity I and V are not affected by rotation; from equations (22.11) we find that, if Q' and U' are the values after rotation,

$$Q' = Q \cos 2\varepsilon + U \sin 2\varepsilon$$
$$U' = -Q \sin 2\varepsilon + U \cos 2\varepsilon$$

(24.14)

If we now change from I,Q to I_r, I_l, we find after some algebra that a rotation through the angle ε is equivalent to multiplying the vector intensity by the following matrix:

$$\underline{R}(\varepsilon) = \begin{pmatrix} \cos^2\varepsilon & \sin^2\varepsilon & \frac{1}{2}\sin 2\varepsilon & 0 \\ \sin^2\varepsilon & \cos^2\varepsilon & -\frac{1}{2}\sin 2\varepsilon & 0 \\ -\sin 2\varepsilon & \sin 2\varepsilon & \cos 2\varepsilon & 0 \\ 0 & 0 & 0 & 1 \end{pmatrix} \qquad (24.15)$$

One can then obtain the Stokes parameters in the meridian plane by using equation (24.13) for ε, and then multiplying the above matrix by the vector intensity of equations (24.8) or (24.9).

The analysis in this section has been simplified in order that the main physical principles will not be obscured. The exact problem of planetary scattering, including the effects of a reflecting ground layer, is solved by Chandrasekhar in Chapter 10 of Radiative Transfer. Van de Hulst also has some interesting scattering problems involving polarization in his book Light Scattering by Small Particles, Wiley, New York, 1957.

Appendix 1: Physical and Astronomical Constants

Speed of light	c	$= 2.9979 \times 10^{10}$ cm sec^{-1}
Boltzmann constant	k	$= 1.3805 \times 10^{-16}$ erg K^{-1}
Planck constant	h	$= 6.6256 \times 10^{-27}$ erg sec
Electron charge	e	$= 4.8030 \times 10^{-10}$ g$^{1/2}$ cm$^{3/2}$ sec^{-1}
Electron mass	m_e	$= 9.1091 \times 10^{-28}$ g
Proton mass	m_p	$= 1.6725 \times 10^{-24}$ g
Mass of unit atomic weight	m_u	$= 1.6604 \times 10^{-24}$ g
Gas constant	R	$= 8.3143 \times 10^{7}$ erg K^{-1}mol^{-1}
Gravitational constant	G	$= 6.670 \times 10^{-8}$ g^{-1}cm^{3}sec^{-2}
Stefan-Boltzmann constant	σ	$= 5.6697 \times 10^{-5}$ erg cm^{-2}sec^{-1}K^{-4}
Bohr radius	a_0	$= 5.2917 \times 10^{-9}$ cm
Avogadro number	N_A	$= 6.0225 \times 10^{23}$ atoms mol^{-1}
Mass of Sun	M	$= 1.989 \times 10^{33}$ g
Radius of Sun	R	$= 6.960 \times 10^{10}$ cm
Luminosity of Sun	L	$= 3.90 \times 10^{33}$ erg sec^{-1}
Effective temperature of Sun	T_e	$= 5800$ K
Surface gravity of Sun	g	$= 2.739 \times 10^{4}$ cm sec^{-2}

1 eV $= 1.60210 \times 10^{-12}$ erg
  $=$ energy of photon of wavelength 1.23981×10^{-4} cm
  $=$ energy of photon of frequency 2.41804×10^{14} sec^{-1}

Appendix 2: Problems

1. A sphere is a black body at temperature T_1. It has radius R and is on the y axis distant r_1 from the origin ($r_1 \gg R$). A cube is a black body of temperature T_2, with edges of length s and being parallel to the coordinate axes. The cube is on the x axis distant r_2 from the origin ($r_2 \gg s$). Find the mean intensity and flux of the combined radiation as observed from the origin.
Answer:

$$J = \frac{1}{4\pi} \frac{\pi R^2}{r_1^2} B(T_1) + \frac{s^2}{r_2^2} B(T_2)$$

$$\vec{F} = -\hat{y} \frac{R^2}{r_1^2} B(T_1) - \hat{x} \frac{s^2}{r_2^2} B(T_2)$$

where $\hat{x}$ and $\hat{y}$ are unit vectors along the coordinate axes.

2. Black body radiation of temperature T_1 passes through a box containing atoms in LTE at temperature T_2. Show that a line formed by the atoms in the box will be in absorption if $T_1 > T_2$ and in emission if $T_1 < T_2$.

3. The mean free path of light photons in interstellar space is about 1000 pc = 3×10^{21} cm. If the mass density of the dust particles causing the absorption is about 10^{-26} g cm^{-3}, find the size and particle density of the dust. Assume the particles are spheres and absorb with their geometric cross section.
Answer: radii of the order of 10^{-5} cm, N about 10^{-12} particles per cm^3.

4. A uniform sphere of radius R has constant absorption and emission coefficients k and j. Find the intensity at an external point distant r from the center of the sphere.

Answer:
$$I_\nu = \frac{j_\nu}{k}(1 - e^{-ks})$$

where

$$s = 2(R^2 - r^2\sin^2\theta)^{1/2}$$

for values of the angle θ less than $\sin^{-1}(R/r)$.

5. Assume that the radius of interaction of neutral hydrogen atoms is a Bohr radius, and that of free protons and electrons is the distance for which the PE of interaction is equal to the mean KE of the particles. Then find the mean free paths for collisions of particles:
    A. in the solar atmosphere with T = 6000 K, $\rho = 10^{-7}$ g cm^{-3}.
    B. in the solar interior with T = 10^7 K, $\rho = 1$ g cm^{-3}.
Answer: A. about .03 cm; B. about 10^{-5} cm.

6. What determines the ratio of the intensities at two different frequencies emerging normal from a semi-infinite atmosphere if
    a) the source function is constant with depth and
        1) absorption is constant with frequency?
        2) absorption varies with frequency?
    b) the source function varies with depth and
        1) absorption is constant with frequency?
        2) absorption varies with frequency?

7. The absorption coefficient in a star is nearly constant with frequency except in a very narrow region. In this region k rises to a sharp maximum, then drops to a sharp minimum, then rises back to its normal value. What does the intensity look like in this region if
    1) the source function increases into the star?
    2) the source function decreases into the star?
Answer: In 1) the intensity fluctuates opposite to k, while in 2) it fluctuates the same way as k.

8. A uniform plane slab has the same absorption coefficient as the star in question 7. Determine the spectrum in the interesting frequency region for various thicknesses of the slab, i.e., from optically thin everywhere to thick everywhere.

9. A uniform slab has the following two absorption and

emission coefficients:

$$j_1 = e_1 \nu^2 \qquad k_1 = a_1(\nu - 100\nu_0) \text{ for } \nu > 100\nu_0$$
$$\qquad\qquad\qquad k_1 = 0 \qquad\qquad\quad \text{ for } \nu < 100\nu_0$$
$$j_2 = e_2 \nu^{-2} \qquad k_2 = a_2 \nu^{-3}$$

The a's, e's, and ν_0 are constants. At the frequency $\nu = \nu_0$, $j_1 = j_2$ and the optical thickness of the slab is 100. Sketch the schematic variation of I_ν with frequency over the entire spectrum.

10. A gray atmosphere in LTE has convection such that the radiative flux is equal to σT_e^4 for optical depths above τ_0, and below this level

$$F_r = \frac{\tau \sigma T_e^4}{\tau_0} \exp{-(\tau/\tau_0 - 1)}$$

Find the temperature distribution using the Eddington approximation. Note that T^4 approaches the limit $T_e^4(1/2 + 9\tau/4)$ at large depths.

11. Outline the steps by which the structure of a model atmosphere could be found if the pressure-density relation were known instead of the temperature-optical depth relation.

12. Show that if the temperature is not too high, a model atmosphere having X, Y, and g for the H abundance, He abundance, and surface gravity is nearly the same as one with X', Y', and gravity $g' = (X'/X)g$.
    Hint: check the effects of varying the parameters on the equation of hydrostatic equilibrium (11.14).

13. Calculate a simplified model of the solar atmosphere by making the following approximations: 1) Assume the Eddington approximation temperature distribution (8.11) with T_e = 5800 K. 2) Assume the Rosseland opacity κ_0 is numerically equal to the optical depth τ_0. 3) Assume something simple concerning abundances, such as 80% neutral H, 15% He all being neutral, and 5% (by mass) heavy elements, these being perhaps 10% ionized. Compare the results of this calculation with Table 5 on page 51.

14. Calculate the spectrum from the model of question

14 as follows: The continuous absorption of H^- is given roughly by the relation

$$\frac{\kappa}{\kappa_o} = 0.0023\lambda'^2 + (-0.92 + 0.54\lambda' - 0.03\lambda'^2)_{>0}$$

where λ' is wavelength in units of 1000A. The quantity in parentheses, representing the bound-free contribution, is to be included only if it is positive.

Ignoring the depth dependences of absorption coefficients, find the temperature corresponding to optical depth unity for a series of wavelengths between 4000A and 20,000A; then use the relation $I_\nu \simeq B_\nu(\tau=1)$ to calculate the spectrum. Compare this with the wavelength dependence of the absorption given above, and also with the Planck function of 5800 K.

15. Describe how the frequency distribution of the emitted radiation would be affected if the amount of convection in a model is changed.
Hint: Regions of the spectrum having low absorption can see the deep convective layers better than those having high absorption.

16. All atoms in a gas are moving with the same speed, but in random directions. Show that the profile function for the Doppler effect has a rectangular profile.

17. Supergiants have much turbulence in their atmospheres along with low pressures; dwarfs have high pressures but little turbulence. Assuming that this turbulence is of the micro-variety (in reality there is also macro), compare strong lines and weak lines in the atmospheres of supergiants and dwarfs of the same temperatures.

18. If a small amount of H in the Sun were replaced by He, how would an absorption line of sodium be affected? Of hydrogen? Of helium?
Hint: The prominence of an absorption line depends mainl on the ratio of line of continuous absorption.

19. Assume that the Planck function varies with the continuum optical depth as

$$B_\nu = B_o(b\tau_c + 1) + Ae^{-\alpha\tau_c}$$

with $\alpha \gg 1$. This form simulates the chromospheric temperature rise. Show that the condition for a very strong line to rise above the continuum at its center is approximately that $A > bB_0$. Assume LTE.

20. The Planck function at a certain frequency is, in arbitrary units, given by

$$B_\nu = 5 \qquad \tau_c < 10^{-3}$$

$$B_\nu = 1 + \tau_c \qquad \tau_c > 10^{-3}$$

Find the approximate quantitative relation between I_ν/I_c and u for an LTE line that is borderline in strength between
a) weak and medium
b) medium and strong.

21. Two lines of an element have the following properties:

	f	λ	g	χ
line 1	0.03	6709A	3	5 eV
line 2	0.09	6730A	10	6 eV

In a star of temperature about 10^4 K, how does the ratio of the equivalent widths vary with the abundance of the element?
Answer: $n_0(2) \simeq 3n_0(1)$. For very low abundances, both lines are weak and $W_2 \simeq 3W_1$. At higher abundances the lines become saturated and $W_1 \to W_2$. As both lines become strong at very large abundances, $W_2 \to 1.7 W_1$.

22. Show from equation (22.6) how measurements of $I(\phi,\varepsilon)$ can be used to distinguish between pure circularly polarized radiation and that that is unpolarized.

23. Show directly from equations (22.1) that $I^2 \geq Q^2 + U^2 + V^2$.

24. Describe how a magnetic plasma as is discussed in Section 23 can emit radiation having non-zero circular polarization even if the emission coefficient in V is zero and if any incident radiation has V = 0.
Hint: The polarized absorption coefficient is an essential part of the answer.

Index

Absorption, 6, 15, 65
Absorption coefficient, 6-8
    continuous, 62-64
    for polarized radiation, 106
    line, 65-68
Allen, C. W., 36
Aller, L. H., xi
Athay, R. G., 18, 74
Blanketing, 62-64
Böhm-Vitense, E. 55
Boltzmann excitation equation, 17
Broadening function, see profile function
Burbidge, E. M., 64
Burbidge, G. R., 64
Carbon, D. F., xi
Chandrasekhar, S., xi, 29-30, 82, 107, 119
Chandrasekhar mean, 49
Collisions, 68
Convection, 52-60
Curve of growth, 95-96
Damping constant, 77, 83, 90-91
Damping wings, 90-92
Degree of polarization, 105-106
Departure coefficient, 67
Doppler broadening, 74-76
Doppler core, 84, 90-92
Doppler width, 75
Eddington approx., 26-29

Effective temperature, 25
Eggen, O. J., 64
Einstein coefficients, 15-18, 65-66
Emission coefficient, 8
    for polarized radiation, 106
    scattering, 13-14
Energy density, 3
Equation of transfer, 9-11
    for polarized radiation, 106
Equivalent width, 85-96
Excitation equation, 17
Excitation temperature, 17
Exponential-integral function, 20-21
Faraday effects, 108
Faraday rotation, 111
Finn, G. D., 84
Flux, 2-3
f-value, 66
Gaussian quadratures, 30-31
Gibson, E. G., xi
Ginzburg, V. L., 107
Gray, D. F., xi
Gray atmosphere, 23-38
Greenstein, J. L., xi
Hummer, D. G., 84
Hydrogen convection zone, 55-56
Hydrostatic equilibrium, 39-43
Impact broadening, 80-83
Induced emission, 15
Intensity, 1
Isotropic scattering, 14-15, 36-38

Jefferies, J. T., xi, 66, 74
Kinetic temperature, 17-18
Kurucz, R. L., 51
Limb darkening, 28
Line absorption, 65-68
Line broadening, 74-84
Line formation, 65-96
Line profile, 85-96
Local thermodynamic equilibrium (LTE), 18-19
Lorentz, H. A., 83
Lorentz profile, 83, 91
Macroturbulence, 75-76, 87-88
Magnetic plasma, transfer in, 107-112
Mark, C., 25
Mean free path, 7
Mean intensity, 2
Microturbulence, 75-76
Mihalas, D., xi, 18, 49, 66, 74
Milne-Eddington approx., 88-96
Model atmospheres, 39-45
Mugglestone, D., 84
Natural broadening, 76-78
Nearest neighbor approx., 81-82
Negative hydrogen ion, 24, 56-57, 62-63
Opacity, 41, 45-49
Optical distance, 11
Oscillator strength, 66
Pacholczyk, A. G., 107, 112
Pagel, B. E. J., 18
Pecker, J.-C., xi, 49
Phase function, 13-14
Phase matrix, 113
Pierce, A. K., 60-61
Planck function, 16
Plasma, transfer in a magnetic, 107-112
Polarization, 97-119
  ellipse, 97-98
Poynting vector, 2
Pressure broadening, 78-84

Profile function, 65-66, 74-84
Radiation pressure, 4, 40-43
Radiative equilibrium, 23
Rayleigh scattering, 14, 112-119
Residual intensity, 87
Resonance broadening, 79
Rosseland mean, 48
Sandage, A. R., 64
Scattering, 6, 8, 12-14
  isotropic, 14-15, 36-38
Semi-empirical models, 60-62
Sky brightness and polarization, 112-119
Source function, 11-19
  line, 67-74
Spontaneous emission, 15
Stark effect, 79
Statistical broadening, 80-82
Stokes, G. G., 100
Stokes parameters, 101
Syrovatskii, S. I., 107
Temperature, effective, 25
  excitation, 17
  kinetic, 17-18
Thermodynamic equilibrium (TE), 16-19
Thomas, R. N., 69
Two-level atom, 68-74
Unsöld, A., xi, 53
Van de Hulst, H. C., 119
Vander Waals broadening, 79
Vitense, E., 55
Voigt profile, 84, 90-91
Waddell, J. H., 60-61
Weisskopf, V., 77
Wigner, E., 77
Wildey, R. L., 64
Wings of lines, 90-92